Nanostructured Materials and Nanotechnology III

T0364395

Nanostructured Materials and Nanotechnology III

*A Collection of Papers Presented at the
33rd International Conference on
Advanced Ceramics and Composites
January 18–23, 2009
Daytona Beach, Florida*

Edited by
Sanjay Mathur
Mrityunjay Singh

Volume Editors
Dileep Singh
Jonathan Salem

A John Wiley & Sons, Inc., Publication

Published by John Wiley & Sons, Inc., Hoboken, New Jersey.
Published simultaneously in Canada.

For general information on our other products and services or for technical support, please contact our
Customer Care Department within the United States at (800) 762-2974, outside the United States at
(317) 572-3993 or fax (317) 572-4002.

Wiley also publishes its books in a variety of electronic formats. Some content that appears in print may
not be available in electronic format. For information about Wiley products, visit our web site at
www.wiley.com.

Library of Congress Cataloging-in-Publication Data is available.

ISBN 978-0-470-45757-3

Printed in the United States of America.

10 9 8 7 6 5 4 3 2 1

Contents

Preface

The 3rd International Symposium on Nanostructured Materials and Nanotechnology was held during the 33rd International Conference on Advanced Ceramics and Composites, in Daytona Beach, Florida during January 19–23, 2009. This symposium provided, for the third consecutive year, an international forum for scientists, engineers, and technologists to discuss new developments in the field of nanotechnology. This year's symposium was dedicated to Prof. Koichi Niihara, Nagaoka University of Technology, Nagaoka, Japan to recognize his seminal contributions in the field of nanocomposite materials. The symposium covered a broad perspective including synthesis, processing, modeling and structure-property correlations in Nanomaterials and nanocomposites. More than 85 contributions (invited talks, oral presentations, and posters) were presented by participants from universities, research institutions, and industry, which offered interdisciplinary presentations and discussions indicating strong scientific and technological interest in the field of nanostructured systems. The geographical spread of the symposium was impressive with participants coming from more than fifteen countries.

This issue contains 16 peer-reviewed (invited and contributed) papers covering various aspects and the latest developments related to processing, modeling and manufacturing technologies of nanoscaled materials including CNT and clay-based composites, nanowire-based sensors, new generation photovoltaic cells, plasma processing of functional thin films, ceramic membranes and self-assembled functional nanostructures. Each manuscript was peer-reviewed using The American Ceramic Society review process.

The editors wish to extend their gratitude and appreciation to all the authors for their cooperation and contributions, to all the participants and session chairs for their time and efforts, and to all the reviewers for their valuable comments and suggestions. Financial support from the Engineering Ceramic Division of The American Ceramic Society is gratefully acknowledged. The invaluable assistance of the staff of the meetings and publication departments of The American Ceramic Society is gratefully acknowledged, which was instrumental in the success of the symposium.

We believe that this issue will serve as a useful reference for the researchers and technologists interested in science and technology of nanostructured materials and devices.

SANJAY MATHUR
University of Cologne
Cologne, Germany

MRITYUNJAY SINGH
Ohio Aerospace Institute
Cleveland, Ohio, USA

Introduction

The theme of international participation continued at the 33rd International Conference on Advanced Ceramics and Composites (ICACC), with over 1000 attendees from 39 countries. China has become a more significant participant in the program with 15 contributed papers and the presentation of the 2009 Engineering Ceramic Division's Bridge Building Award lecture. The 2009 meeting was organized in conjunction with the Electronics Division and the Nuclear and Environmental Technology Division.

Energy related themes were a mainstay, with symposia on nuclear energy, solid oxide fuel cells, materials for thermal-to-electric energy conversion, and thermal barrier coatings participating along with the traditional themes of armor, mechanical properties, and porous ceramics. Newer themes included nano-structured materials, advanced manufacturing, and bioceramics. Once again the conference included topics ranging from ceramic nanomaterials to structural reliability of ceramic components, demonstrating the linkage between materials science developments at the atomic level and macro-level structural applications. Symposium on Nanostructured Materials and Nanocomposites was held in honor of Prof. Koichi Niihara and recognized the significant contributions made by him. The conference was organized into the following symposia and focused sessions:

Symposium 1	Mechanical Behavior and Performance of Ceramics and Composites
Symposium 2	Advanced Ceramic Coatings for Structural, Environmental, and Functional Applications
Symposium 3	6th International Symposium on Solid Oxide Fuel Cells (SOFC): Materials, Science, and Technology
Symposium 4	Armor Ceramics
Symposium 5	Next Generation Bioceramics
Symposium 6	Key Materials and Technologies for Efficient Direct Thermal-to-Electrical Conversion
Symposium 7	3rd International Symposium on Nanostructured Materials and Nanocomposites: In Honor of Professor Koichi Niihara
Symposium 8	3rd International symposium on Advanced Processing & Manufacturing Technologies (APMT) for Structural & Multifunctional Materials and Systems

Symposium 9	Porous Ceramics: Novel Developments and Applications
Symposium 10	International Symposium on Silicon Carbide and Carbon-Based Materials for Fusion and Advanced Nuclear Energy Applications
Symposium 11	Symposium on Advanced Dielectrics, Piezoelectric, Ferroelectric, and Multiferroic Materials
Focused Session 1	Geopolymers and other Inorganic Polymers
Focused Session 2	Materials for Solid State Lighting
Focused Session 3	Advanced Sensor Technology for High-Temperature Applications
Focused Session 4	Processing and Properties of Nuclear Fuels and Wastes

The conference proceedings compiles peer reviewed papers from the above symposia and focused sessions into 9 issues of the 2009 Ceramic Engineering & Science Proceedings (CESP); Volume 30, Issues 2-10, 2009 as outlined below:

- Mechanical Properties and Performance of Engineering Ceramics and Composites IV, CESP Volume 30, Issue 2 (includes papers from Symp. 1 and FS 1)
- Advanced Ceramic Coatings and Interfaces IV Volume 30, Issue 3 (includes papers from Symp. 2)
- Advances in Solid Oxide Fuel Cells V, CESP Volume 30, Issue 4 (includes papers from Symp. 3)
- Advances in Ceramic Armor V, CESP Volume 30, Issue 5 (includes papers from Symp. 4)
- Advances in Bioceramics and Porous Ceramics II, CESP Volume 30, Issue 6 (includes papers from Symp. 5 and Symp. 9)
- Nanostructured Materials and Nanotechnology III, CESP Volume 30, Issue 7 (includes papers from Symp. 7)
- Advanced Processing and Manufacturing Technologies for Structural and Multifunctional Materials III, CESP Volume 30, Issue 8 (includes papers from Symp. 8)
- Advances in Electronic Ceramics II, CESP Volume 30, Issue 9 (includes papers from Symp. 11, Symp. 6, FS 2 and FS 3)
- Ceramics in Nuclear Applications, CESP Volume 30, Issue 10 (includes papers from Symp. 10 and FS 4)

The organization of the Daytona Beach meeting and the publication of these proceedings were possible thanks to the professional staff of The American Ceramic Society (ACerS) and the tireless dedication of the many members of the ACerS Engineering Ceramics, Nuclear & Environmental Technology and Electronics Divisions. We would especially like to express our sincere thanks to the symposia organizers, session chairs, presenters and conference attendees, for their efforts and enthusiastic participation in the vibrant and cutting-edge conference.

DILEEP SINGH and JONATHAN SALEM
Volume Editors

NANOWIRES AS BUILDING BLOCKS OF NEW DEVICES: PRESENT STATE AND PROSPECTS

F. Hernandez-Ramirez[1,2], J. D. Prades[2], R. Rodriguez-Diaz[2], A. Romano-Rodriguez[2], J. R. Morante[2,3], S. Mathur[4]

[1] Electronic Nanosystems S.L, Barcelona, Spain
[2] EME/XaRMAE/IN²UB, Dept. d'Electrònica, Universitat de Barcelona, Barcelona, Spain
[3] IREC, Catalonia Institute for Energy Research, Barcelona, Spain
[4] Department of Inorganic Chemistry, University of Cologne, Cologne, Germany

ABSTRACT

Single-crystalline semiconductor nanowires have emerged as potential building blocks of new devices and circuit architectures due to their astonishing performance derived from their reduced size and well-controlled chemical and physical properties. To date, a great effort has been made in the synthesis and the electrical characterization of these nanomaterials. Nevertheless, the development of real proof-of-concepts nanodevices is still in the preliminary stages. In this contribution, the use of individual metal-oxide nanowires to obtain competitive devices and their integration in portable platforms will be reviewed. Nanowires, which are among the most promising nanomaterials, have demonstrated their suitability to fabricate gas sensors or photodetectors with high sensitivity, stable and reproducible characteristics; and they are generally considered excellent candidates to study some of the phenomena arising at the nanoscale. Finally, future work is outlined which would provide both more complex nanowire based circuit architectures as well as methods for producing these structures in a commercializable fashion.

INTRODUCTION

Monocrystalline semiconductor nanowires have unique properties derived from their high surface-to-volume ratio and well-defined atomic arrangement [1]. This facilitates their integration in many different individual nanowire-based devices such as gas sensors or photodetectors [1-4], overcoming some of the major technological drawbacks which are commonly found in their thin-film counterparts (i.e. high power consumption, lack of stability and drift problems). Nevertheless, the development of competitive devices based on nanomaterials is still in the preliminary stages, and the launch of commercial products remains a major and unsolved challenge due to the difficulties in electrically contacting nanowires and making the best use of their full potential [5]. In this paper, the ultimate advantages of using individual metal-oxide nanowires as building blocks of advanced functional devices are briefly surveyed, and the first portable prototypes based on them are shown. Finally, a discussion on current challenges for the production of commercializable nanowire based complex circuit architectures will be outlined.

ON THE ADVANTAGES OF USING NANOWIRES IN FUNCTIONAL DEVICES

The grounds of many metal-oxide devices are based on the chemico-electrical transduction reactions which take place at their surfaces [4, 6]. For this reason, increasing the surface-to-volume ratio is considered the best strategy to maximize their response towards different external stimuli such as gas molecules or impinging photons. Individual single-crystalline nanowires meet this condition, and thanks to their low mass they are excellent candidates to be integrated in a new generation of low-consumption devices [7]. Nevertheless, the controlled manipulation of individual nanowires is by no means a straightforward process, making necessary the development of well-controlled and advanced nanofabrication techniques [8]. To circumvent this technological obstacle, most of the overwhelming number of articles on metal oxide nanowires published up to now is based on the use of bundles of nanowires instead of individual ones. However, the typical drawbacks of thin-film metal-oxide devices are also found in the characterization of multiple nanowires (i.e. parasitic electrical contributions and stability problems). Figure 1.a shows a diagram of a thin-film metal-oxide device formed by a large amount of nanoparticles put together onto a supporting substrate. Herein grain boundaries work as Schottky barriers that represent the main contribution to the overall electrical resistance [5]. The spread in both the size and the intrinsic properties of these nanoparticles complicates the interpretation of fundamental studies on the physical and chemical reactions which determine their performance, and the complexity of the inter and intragrain electrical transport is considered a major obstacle to improve present metal-oxide devices and technologies. It is noteworthy that the same conclusions are reached if multiple-nanowire devices are characterized, since the experimental set-up is similar to the former description, just as Figure 1 shows. In fact and from a theoretical point of view, there are not significant differences in using bundles of nanowires instead of conventional thin-films to fabricate functional devices with metal-oxides.

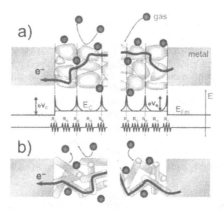

Fig. 1 Schematic diagrams of different types of conductometric gas sensors based on metal oxides. (a) Commercial thin-film sensor formed by a layer of nanoparticles. Here, electrons must go through a network of nanocrystals with different size and shape. From an energy point of view, electrons are to overcome potential barriers [(i) metal-semiconductor barriers (eV_C) and (ii) intergrain boundary barriers (eV_B)]. The overall influence of the exposure gas on the height of the barriers determines the final response of the sensor. This is equivalent to a network of resistors [(i) metal-semiconductor contacts (R_C), (ii) grain boundary interfaces (R_B) and (iii) metal-oxide grains (R_G)]. (b) Multi-nanowire sensor. The above mentioned discussion is valid here as well

On the contrary, the development of devices based on individual nanowires has many advantages. Nanowires, which are usually described as pure resistors R_{NW} (Fig.2) [9], have well-defined surfaces and exhibit good electrical and physical stability as function of time, enabling the fabrication of lasting devices. Furthermore, the complete absence of nanowire-nanowire boundaries eliminates one of the major sources of drift and aging of the devices [10]. In fact, these two problems typical of metal-oxides are usually attributed to the aforementioned grain boundary effects [10].

In short, individual nanowires, whose electrical properties and responses towards different chemico-physical stimuli can be determined by fixing their dimensions and intrinsic parameters such as the free carrier concentration n_d and mobility μ [10, 11], are considered excellent building blocks to fabricate new semiconductor devices with well-defined properties.

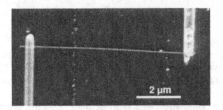

Fig. 2 SnO₂ nanowire electrically contacted with FIB nanolithography techniques.

ON THE FIRST PROTOTYPES USING INDIVIDUAL NANOWIRES

To date, the practical fabrication of complex circuit architectures based on individual nanowires remains an unattainable objective. Nevertheless, many of the theoretical potentials of nanowires have already been demonstrated in simpler devices, for instance their use as gas sensors or UV photodetectors are extensively reported elsewhere [10-14].

The outstanding responses towards both reducing (CO) and oxidizing (NO_2) gas species of individual SnO_2 nanowires confirmed most of the advantages outlined in the former section (Figure 3). Moreover, the detailed analysis of experimental data allowed the development of a theoretical model able to describe the gas sensing mechanisms in a simple way.

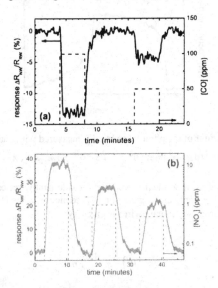

Fig. 3 (a) Response of a SnO_2 nanowire towards different CO pulses at T = 573 K. (b) Response of a SnO_2 nanowire towards NO_2 pulses at T = 448 K. In both cases, fast and reproducible behaviors are monitored, with excellent recovery of the synthetic air baseline resistance

While interaction mechanisms between gas molecules and nanowires are extremely complex; the sensing principle can be described by pure surface effects [10, 11]. To a rough approximation, adsorption of gas molecules at the nanowire modulates the width of a depleted region close to the external shell. This modifies the conduction channel through it and as a consequence R_{NW}. According to this assumption, R_{NW} under exposure to gas is given by:

$$R_{NW} = \frac{\rho L}{\pi(r - \lambda)^2} \tag{1}$$

where ρ is the nanowire resistivity, L the nanowire's length, r the nanowire's radius and λ the width of the depletion layer created by adsorbed molecules. Equation 1 lays down a connection between the nanowire's radius and the gas response: the thinner the nanowire is, the higher the gas response [10].

To meet this requirement, nanowires with radii below r = 40 nm are commonly used for this purpose, giving rise to technological issues derived from working at the nanoscale such as high contact resistance at the metal-nanowire interfaces [9]. Nevertheless, most of these problems are circumvented with different operating strategies reported elsewhere [9], paving the way to better gas nanosensors than their thin-film counterparts.

On the other hand and as far as the use of nanowires as photodetectors is concerned, Prades et al. demonstrated that individual SnO_2 and ZnO nanowires exhibit outstanding response towards impinging UV photons, and determined systematic fabrication strategies to enhance their responses.

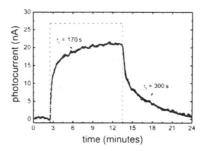

Figure 4. Dynamic behaviour of the photoresponse I_{ph} measured with one single ZnO nanowire when a UV pulse is applied (dashed line) ($\Phi_{ph} = 3.3 \cdot 10^{18} \, ph \, m^{-2} s^{-1}$; $\lambda = 340 \pm 10 \, nm$; $V = 1V$).

Photodetectors based on individual n-type metal-oxide nanowires can be studied using the fundamental principles ruling light carrier generation on semiconductors [14]. Thus and according to this assumption, photocurrent I_{ph} in nanowires is given by

$$I_{ph} = j_{ph} \left(\alpha^{-1} W \right) = q \, \frac{W}{L} \, \beta \eta \tau \mu^* \, V \Phi_{ph} \tag{2}$$

where three different contributions are clearly identified. The first one is related to geometric parameters of the device (W/L), the second one to the intrinsic properties of nanowires ($\beta \eta \tau \mu^*$) and the third one only depends on the working conditions ($V \Phi_{ph}$) [14]. Here, j_{ph} is the current density, α is the absorption profile of the material, W is the width of the photodetector, L is its length, q is the fundamental electrical charge, β is the fraction of photons not reflected by the surface, η is the quantum efficiency of pairs generations by one photon, τ is the carrier lifetime, μ^* is the effective carrier mobility, V is the applied voltage and Φ_{ph} is the photon flux.

As far as the geometry of photodetectors is concerned, it is clear from Eq.2 that I_{ph} is enhanced by increasing the width (W) of the photoactive area. A convenient way to reach this objective is electrically contacting several nanowires in parallel [14]. On the other hand, distance between contacts (L) also determines the response of the photodetector (see Eq.2). L not only influences the photocaptured area ($W \cdot L$) but also determines the effective electric field E inside the nanowire due to the bias voltage V applied externally. Indeed, this second aspect dominates the overall contribution of L to the photoresponse. Therefore, it can be concluded that higher-gain photodetectors are obtained by diminishing this parameter [14]. The lower limit for L will strictly depend on the precision of the nanolithography technique we use to fabricate the electrical contacts.

In short, it can be asserted that single-nanowire prototypes are extremely useful to demonstrate the potential of nanowires. Nevertheless, the present results and responses can only be improved if complex architectures, such as multiple-nanowires contacted in parallel, are fabricated and characterized. For this reason, many research efforts are currently been devoted to reach this ambitious goal.

TOWARDS REAL DEVICES

The possibility of monitoring the electrical properties of individual nanowires with portable cost-effective and consumer-class electronics (Fig.5) was recently demonstrated [12]. These low-cost instruments, compared to lab equipments, were able to detect and quantify the response of individual nanowires towards UV light pulses and different gases with long-term stability thanks to the low current injected by the platform to the nanowire [12].

Figure 5. Low-cost electronic platform designed to characterize nanowires

It is well-known that metal oxide materials need to be heated at a specific temperature to maximize their response to a specific target [15]. Therefore, the use of a heater becomes a necessary tool to modulate the final performance of these materials. To solve this issue, both bottom-up and top-down fabrication techniques have been successfully integrated in a single process; nanowires are electrically contacted to a micro-hotplate with an integrated heater [12] (Fig.6).

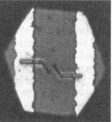

Figure 6. (right) Individual SnO$_2$ nanowire electrically contacted to two platinum electrodes. (left) Interdigitated platinum electrodes surrounded by a microheater. The nanowire is located in the middle (red square)

This set up allows modulating the effective temperature of the wire as function of the power dissipated at the heater in a fast and completely reproducible manner. It is noteworthy that this solution combined with a good electronic interface, which integrates the thermal control of the nanowire, is extremely useful in many sensing applications [12]. Other architectures are also being explored to solve one of the major issues of sensors: the lack of selectivity to interfering stimuli. Typical examples are photodetectors sensible to a wide range of light wavelengths or gas sensors responsive to parasitic species, such as moisture. For this reason, brand-new studies are attempting to develop electronic systems based on arrays of different individual metal oxide nanowires [16]. According to this approach, their responses are monitored in parallel, and the specific sensing characteristics of each one are determined and electronically recorded following one of the strategies previously described. Later, the data are processed by pattern recognition software to determine the composition of the external stimuli [17]. Although these studies are currently ongoing, they are the most promising solution to overcome the lack of selectivity, which is characteristic of metal oxide nanowires.

FUTURE CHALLENGES

Despite the advances in nanolithography techniques which have made possible the fabrication of devices based on individual nanowires, these techniques are only suitable for prototyping and academic purposes. To extend the use of nanowires to low cost and large scale fabrication processes, self-assembling techniques must be taken into account. In this direction, the first steps had been made to self-align one-dimensional metal-oxide nanostructures by means of dielectrophoretic techniques [18-21].

Dielectrophoresis is an attractive alternative for the positioning and alignment of nanowires thanks to its low-cost, simplicity and flexibility [18, 22]. This method is based on the well-known forces that appear when dielectrically polarized materials are in a medium in which a non uniform electric field is applied [23,24]. It has been tested for different nanomaterials like single- [25-27] and multi-walled carbon nanotubes [77, 78], polymeric [79], metal [80, 81] and semiconductor nanowires [19, 22], and of course metal oxide nanowires [18-21].

Dielectrophoresis can be applied to the fabrication of a new generation of nanodevices and it can be easily combined with other techniques like e-beam or conventional photolithography. If the right design of electrodes is used, nanowires are not only aligned but also positioned at any desired position, and thus, the time necessary to fabricate a device is significantly reduced. This advantage can be applied to the fabrication of the simplest electronic elements, like rectifying junctions [32] and transistors [26], paving the way for the development of novel electronic devices exclusively based on nanostructured semiconductors materials.

Nevertheless, before reaching this high control of self-assembly techniques, the performance of hybrid designs which combine conventional components integrated in silicon and nanowires devices must be investigated in order to achieve new microsystems with enhanced capacities [33].

CONCLUSIONS

Semiconductor nanowires have novel properties derived from their reduced dimensions and excellent crystallinity, which can be used to obtain functional devices such as gas sensors and UV photodetectors better than their bulk-counterparts. Up to now, simple prototypes based on few nanowires have been fabricated and studied. Nevertheless, complex device architectures remain unattainable objective due to the difficulties of working at the nanoscale in a controlled way. For this reason, self-assembly techniques (i.e. dielectrophoresis) or other alternatives such as electrospinning are considered excellent fabrication alternative to overcome this technological bottleneck thus enabling the development of nanowire based commercial devices in the future.

ACKNOWLEDGMENTS

This work was partially supported by the Spanish Government [projects N – MOSEN (MAT2007-66741-C02-01), and MAGASENS], the UE [project NAWACS (NAN2006-28568-E), the Human Potential Program, Access to Research Infrastructures]. JDP and RJD are indebted to the MEC for the FPU grant. Thanks are due to the European Aeronautic Defense and Space Company (EADS N.V.) for supplying the suspended micromembranes. SM thanks the University of Cologne for financial support.

REFERENCES

[1] M. Law, J. Goldberger and P. Yang, Annu. Rev. Mater. Res., 2004, 34, 83.

[2] S. V. N. T. Kuchibhatla, A. S. Karaoti, D. Bera and S. Seal, Progress Mater. Sci., 2007, 52, 699.

[3] G. Eranna, B. C. Joshi, D. P. Runthala and R. P. Gupta, Critic. Rev. Sol. State Mater. Sci., 2004, 29, 188.

[4] G. Korotcenkov, Sens. Actuators B: Chem., 2005, 107, 209

[5] F. Hernandez-Ramirez, J. D. Prades, R. Jimenez-Diaz, A. Romano-Rodriguez, T. Fischer, S. Mathur, J. R. Morante. Phys. Chem. Chem. Phys. (submitted)

[6] A. Kolmakov, D. O. Klenov, Y. Lilach, S. Stemmer and M. Moskovits, Nano Lett., 2005, 5, 667.

[7] J. D. Prades, R. Jimenez-Diaz , F. Hernandez-Ramirez, S. Barth, J. Pan, A. Cirera, A. Romano-Rodriguez, S. Mathur and J. R. Morante, Appl. Phys. Lett 93, 123110 , 2008

[8] F. Hernandez-Ramirez, A. Tarancon, O. Casals, J. Rodríguez, A. Romano-Rodriguez, J. R. Morante, S. Barth, S. Mathur, T. Y. Choi, D. Poulikakos, V. Callegari and P. M. Nellen, Nanotechnol., 2006, 17, 5577.

[9] F. Hernandez-Ramirez, A. Tarancon, O. Casals, E. Pellicer, J. Rodriguez, A. Romano-Rodriguez, J. R. Morante, S. Barth, S, Mathur. Phys. Rev. B 76, 085429, 2007.

[10] F. Hernandez-Ramirez, J. D. Prades, A. Tarancon., S. Barth, O. Casals, R. Jimenez-Diaz, E. Pellicer, J. Rodriguez, J. R. Morante, M. A. Juli, S. Mathur and A. Romano-Rodriguez, Adv. Funct. Mater., 2008, 18, 2990.

[11] F. Hernández-Ramírez, A. Tarancón, O. Casals, J. Arbiol, A. Romano-Rodríguez and J. R. Morante, Sens. Actuators B: Chem., 2007, 121, 3.

[12] F. Hernandez-Ramirez, J. D. Prades, A. Tarancon, S. Barth, O. Casals, R. Jimenez-Diaz, E. Pellicer, J. Rodríguez, M. A. Juli, A. Romano-Rodriguez, J. R. Morante, S. Mathur, A. Helwig, J. Spannhake and G. Mueller, Nanotechnol., 2007, 18, 495501.

[13] J. D. Prades, F. Hernandez-Ramirez, R. Jimenez-Diaz, M. Manzanares, T. Andreu, A. Cirera, A. Romano-Rodriguez and J. R. Morante, Nanotechnol., 2007, 19, 465501002E

[14] J. D. Prades, R. Jimenez-Diaz, F. Hernandez-Ramirez, L. Fernandez-Romero, T. Andreu, A. Cirera, A. Romano-Rodriguez, A. Cornet, J. R. Morante, S. Barth and S. Mathur, J. Phys. Chem. C, 2008, 112, 14639.

[15] N. Barsan, D. Koziej, and U. Weimar. *Sensors and Actuators B.* 121, 18 (2007).

[16] V. V. Sysoev, B. K. Button, K. Wepsiec, S. Dmitriev, and A. Kolmakov. *Nano Lett.* 6, 8, 1584 (2006).

[17] V. V. Sysoev, J. Goschnick, T. Schneider, E. Strelcov, and A. Kolmakov. *Nano Lett.* 7, 10, 3182 (2007).

[18] S.-Y. Lee, A. Umar, D.-I. Suh, J.-E Park, Y.-B. Hahn, J.-Y. Ahn, S.-K. Lee. *Physica E.* 40, 866 (2008).

[19] H. W. Seo, C.-S. Han, S. O. Hwang and J. Park. *Nanotechnology* 17, 3388 (2006).

[20] S. Kumar, S. Rajaraman, R. A. Gerhardt, Z. L. Wang, P. J. Hesketh. *Electrochimica Acta.* 51, 943 (2005).

[21] J. Suehiro, N. Nakagawa, S. Hidaka, M. Ueda, K. Imasaka, M. Higashihata,T. Okada, M. Hara. *Nanotechnology.* 17, 2567 (2006).

[22] X. Duan, Y. Huang, Y. Cui, J. Wang and C. M. Lieber. *Nature* 409, 66 (2001).

[23] H. A. Pohl. *J. Appl. Phys.* 22 869 (1951).

[24] H. A. Pohl. *Dielectrophoresis.* Cambridge University Press, London, (1978).

[25] M. Lucci, R. Regoliosi, A. Reale, A. Di Carlo, S. Orlanducci, E. Tamburri, ML. Terranova, P.. Lugli, C. Di Natale, A. D'Amico, R. Paolesse. *Sensors and Actuators B* 111, 181 (2005).

[26] P. Stokes and S. I Khondaker. *Nanotechnology* 19, 175202 (2008).

[27] H.W. Seo, C.-S. Han, W, S. Jang , J. Park. *Current Applied Physics* 6, 216 (2006).

[28] S. Tung, H. Rokadia, W. J. Li. *Sensors and Actuators A* 133, 431 (2007).

[29] Y. Dan, Y. Cao, T. E. Mallouk, A. T. Johnson, S. Evoy. *Sensors and Actuators B* 125, 55 (2007).

[30] P. A. Smith, C. D. Nordquist, T. N. Jackson, T. S. Mayer, B. R. Martin, J. Mbindyo, and T. E. Mallouk. *Appl. Phys. Lett.* 77, 1399 (2000)

[31] J. J. Boote and S. D. Evans, *Nanotechnology* 16, 1500 (2005).

[32] Y. Lee, T.-H. Kim, D.-I. Suh, J.-E. Park, J.-H. Kim, C.-J. Youn, B.-K. Ahn, S.-K. Lee. Physica E. 36, 194 (2007).

[33] Evoy, N. DiLello, V. Deshpande, A. Narayanan, H. Liu, M. Riegelman, B.R. Martin, B. Hailer, J.-C. Bradley, W. Weiss, T.S. Mayer, Y. Gogotsi, H.H. Bau, T.E. Mallouk, S. Raman. Microelectronic Engineering. 75, 31 (2004).

MECHANISTIC STUDIES ON CHEMICAL VAPOR DEPOSITION GROWN TIN OXIDE NANOWIRES

Jun Pan, Lisong Xiao, Hao Shen and Sanjay Mathur
Institute of Inorganic Chemstry
University of Cologne
50939 Cologne
Germany

ABSTRACT

Tin oxide nanowire arrays on titania (001) have been successfully fabricated by using of CVD of Sn(O^{t}Bu)$_4$ precursor. The diameter/length of ordered nanowires were well controlled by variation of precursor/substrate temperatures and size of gold catalyst. The morphologies and structures were analyzed by SEM/EDX, HR-TEM and XRD. The wires prefer [101], [-101], [011] and [0-11] growth direction due to the lowest surface energy. The cross-sectional TEM analysis has shown the evidence of epitaxial growth from titania substrate via epilayer to ordered tin oxide nanowires. A model for the nanowire morphology based upon crystallographic relation, defect and preferential growth direction is proposed. The tin arrays can be potentially used as diameter- and shape-dependent sensing unit for detection of gas and bio-molecules.

INTRODUCTION

Non-carbon one-dimensional (1D) nanoscale materials, such as nanotubes (NTs), nanowires (NWs), and nanobelts (NBs), have attracted significant attention due to their unique size- and dimension-dependent electrical, optical, chemical and mechanical properties and possible applications as interconnection and functional components in designing nano-sized electronic and optical devices.[1] Semiconductor metal oxide nanowires display interesting fundamental importance and the wide range of their potential applications in nanodevices.

It is fact that the development of 1D nanowires is in rapid velocity but the satisfied synthetic methods are limited to vapor transport or related methods.[1] The simplest vapor-liquid-solid (VLS) growth mechanism is relatively complicated although the primary phyiscal picture has been recently explored by Ross's group by using in-situ TEM technique.[2] Since the physical alloying, diffusion and necleation are thermodynamic dependent, the chemical control can maybe suppress several unnecessary reactions (kinetic control) resulting in high quality nanowires with better functions. The chemical nanotechnology has demonstrated the great potential in the rational synthesis of nanomaterials in nanoparticles (0D) and films (2D) form.[3] The anisotropic growth of nanomaterials (1D) can be easily realized by modifying the growth facets in order to reduce the energy barrier of growth (nucleation) to stimulate the high growth rate in competing to other crystallographic orientations. A new balance for chemical control over the functional nanowires is required. However, only few reports are focusing on the chemical controlled synthesis of 1D nanostructures.[1]

As a n-type direct wide band semiconductor (E$_g$ = 3.6 eV at 300 K), SnO$_2$ is a key functional material that has been used extensively for gas sensor,[4] optoelectronic devices,[5] catalyst supports,[6] transparent conducting electrodes and antireflective coatings, while responding to the test gas, which effectuates the differential sheath conductivity as a function of the analyte gas concentration.[7]

One-dimensional tin oxide nanostructures have been synthesized by a variety of techniques including vapor transport,[8] carbothermal reduction,[9] molecule-based chemical vapor depostion (MB-CVD),[10] laser ablation of pure tin in an oxidizing Ar/O_2 atmosphere,[11] oxidation of electrodeposited tin wires,[12] oxidation of tin vapors at elevated temperatures,[13] solvothermal synthesis,[14] and electrospinning.[15] To date, many studies have focused on controlled synthesis of metal oxide NWs using the VLS mechanism,[16,17] as well as thermodynamic and kinetic size limit of NWs growth. However, few studies of metal oxide NWs growth kinetics are performed, especially theoretical studies.[18,19]

Molecule-based chemical vapor deposition is a powerful technique for the synthesis of inorganic nanostructures via the well-known vapor-liquid-solid (VLS) growth mechanism because it allows to control the gas phase supersaturation in the CVD reactor. We report here on the grown of tin oxide nanowire by the decomposition of a single molecular precursor $[Sn(O^tBu)_4]$ on gold nanoparticles, on Si, Al_2O_3 and TiO_2 substrates. The main features of MB-CVD approach are that the structural properties of obtained tin oxide NWs can be tuned by modification of molecular precursor (internal) and the regulation of process parameters (external). In this work we will present the controlled synthesis and growth mechanism of ordered tin oxide nanowires in order to show the great potential of the MB-CVD method. Since the ordered nanowires have defined growth direction and surface states, they are suitable to be used in large scale device fabrication.

EXPERIMENTAL

The synthesis of $Sn(O^tBu)_4$ precursor was performed in a modified Schlenk type vacuum assembly, taking stringent precautions against atmospheric moisture. The precursor $Sn(O^tBu)_4$ was synthesized following published procedures. Precursor was purified by distillation or sublimation under reduced pressure (10^{-2} Torr).

A cold-wall horizontal CVD reactor operating under reduced pressure (10^{-6} Torr) was used to investigate the gas-phase decomposition of the tin alkoxide precursor. The TiO_2 (001) and polystalline Al_2O_3 substrates were placed in the quartz reactor on a custom-made graphite susceptor and inductively heated using a 1.5 kW, 350 kHz high frequency generator. The deposition temperature was set in the range 650-750 °C and monitored using a thermocouple attached to the susceptor. The precursor, $Sn(O^tBu)_4$, was introduced in the reactor through a glass flange by applying dynamic vacuum and maintaining the precursor reservoir at the desired temperatures (20 °C). For the deposition of SnO_2 nanowires, TiO_2 and Al_2O_3 substrates were partially coated with thin (2-3 nm) gold films to induce the one-dimensional growth following the vapor-liquid-solid growth mechanism.

The synthesized tin oxide nanowires were characterized by a JSM-7000F (JEOL) scanning electron microscopy (SEM), and a Philips 200 FEG (200 KV) high-resolution transmission electron microscopy (HRTEM). Room temperature X-ray diffraction was performed with a Siemens D-500 diffractometer using Cu-K_α radiation to obtain the structural information of 1D nanostructures.

RESULTS AND DISCUSSION

Figures 1(a) and (b) show typical SEM images of the samples in random and ordered forms, respectively. As in-plane view of the tin oxide NWs confirmed, an oriented growth with a meshlike network of straight NWs of regular dimensions on single crystalline TiO_2 (001) can be observed as shown in Figure 1b. Figure 1a shows the morphology of SnO_2 nanowires on Al_2O_3 polycrystalline substrates. There ultra-long SnO_2 nanowires have lengths of almost 50 micrometers and appear to grow randomly on the Al_2O_3 substrates. The polyhedron particle located on the nanowire tip is identified to be Au by energy dispersive spectroscopy (EDS), indicating vapor-liquid-solid (VLS) growth[20] of the nanowires. The spherical droplets at the tip of the nanowires are commonly considered to be the evidence of the VLS mechanism. The different morphologies on the TiO_2 and Al_2O_3 substrates are attributed to the driving force of crystallographic orientation.

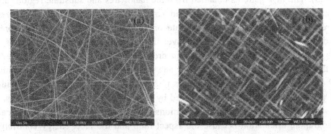

Figure 1. SEM images of SnO_2 nanowires in (a) random and (b) ordered forms.

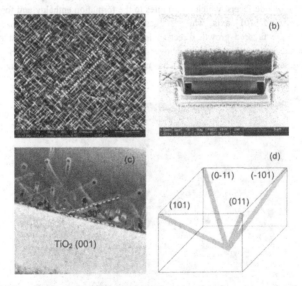

Figure 2. (a - c) Electron microscopic images of FIB preparation for cross-sectional analysis of SnO_2 nanowires on TiO_2 (001) substrate. (d) Four equivalent growth directions of SnO_2 nanowires on TiO_2 (001) substrate.

For ordered SnO_2/TiO_2 system, Figure 2 (a-c) shows electron microscopic images of focused ion beam (FIB) preparation for cross-sectional analysis. It revealed a characteristic angular orientation of the NWs with respect to the substrate, presumably imposed by the fourfold axis of symmetry present along the c axis of the TiO$_2$ (001) substrate, which offered the growing tin oxide nanowires four equivalent growth directions ([011], [0-11], [101] and [-101]) due to mirror images of the unit cell (Fig. 2d). The nanowires are tilted at ca. 30° to the surface of substrate. High-resolution TEM images of ordered tin oxide nanowires revealed that the epitaxial growth of SnO$_2$ nanowires on TiO$_2$ (001) substrates (Fig. 3). The FFT patterns obtained from the nanowires and substrate regions are similar in point forms supporting the epitaxial growth and nature of single crystalline nanowires. Previous research has reported that the choice of substrate orientation and NWs dimensions predominantly influence the degree of alignment and control over the growth axis, with contributions of the bulk energy of tin oxide, the bulk energy of the liquid droplet, the interfacial tension of the liquid–solid interface and the surface tensions of the droplet and the nanowire, respectively.[21] In this system, it showed that the energetically favored growth axis to be along the [101] direction, as evidenced from the sharp FFT pattern (inset, Fig. 3) and confirmed by HR-TEM analysis of several NWs. Apparently the initial growth is governed by the epitaxial relationship, with the favored [001] growth axis dictated by the crystallographic orientation because the aspect ratio and the facet energy contribution is low. With increasing aspect ratio, surface energy dominates the overall growth process resulting in preferential growth in the [101] direction at the expense of [001], possibly a consequence of the growing significance of side facets, which contributes to the formation enthalpy and forces the wire to grow along the observed [101] axis, which suggests that crystal symmetry loses against surface energy.[22] Generally, the favored growth direction is found to be along the lowest surface energy direction in the cubic lattice. Based on the above structural analysis, a purposed growth model is presented here (Fig. 4). It is evident that {011} has more closed packing structure than that of {001} plane.

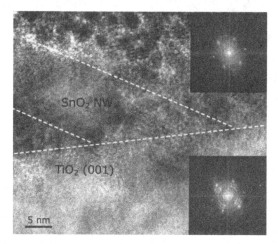

Figure 3. The cross-sectional HR-TEM images showed the epitaxial growth of SnO$_2$ nanowires on TiO$_2$ (001) substrates.

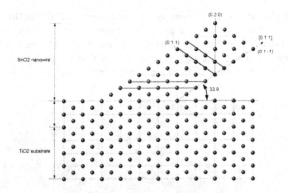

Figure 4. Schematic growth model of ordered tin oxide nanowires on TiO_2 (001) substrate.

A change to the epitaxial growth direction due to higher lattice mismatch can accommodate the strain between the deposited material and the substrate.[23] Hence, the lattice mismatch is a definitive factor to growth direction. In this case, TiO_2 substrate is isostructural with the rutile phase of tetragonal tin oxide. The lattice mismatch is so low (($a_{SnO2} - a_{TiO2}$) / a_{TiO2} = 2.6 % and ($c_{SnO2} - c_{TiO2}$) / c_{TiO2} = 7.7 %) that we can neglect the influence on the growth direction. Hence, the amount of activation energy E_a increases in direct proportion to the lattice mismatch. Due to the nano-scaled dimension of nanowires, the lattice induced strain in the SnO_2/TiO_2 interface can be compensated by structural relaxation. Usually few structural defects in nanowires can be observed. The growth model can also explain the XRD investigation which only (001) reflex of SnO_2 can be observed. Since the {011} nanowires have discrete nano-scaled {001} plane, the (001) reflex can be easily detected. The calculated tilted angle of [011] (ca. 33.9°) is similar to the experimental result (Fig. 2c).

In order to explore the growth mechanism of ordered tin oxide nanowires on TiO_2 (001) substrate, the substrate was partially coated with catalytic Au film (only the circle region) before CVD experiment. Figure 5 shows the different growth of tin oxides on TiO_2 (001) with and without Au catalyst. The enlarged SEM image shows the evolution of SnO_2 from ordered nanowire to nanoparticle. Due to the catalytic role of Au particles, the growth rate of nanowires in circle region is much higher than that of nanoparticles/film in the region outside the circle. The combination of Au catalysts and TiO_2 crystallographic force drives the growth orientation of SnO_2 nanowires whereas only SnO_2 nanoparticles are formed in the uncoated TiO_2 region. The experiment has also shown the surface migration of Au catalyst which the interface has a gradient growth zone (length ~ several micrometer).

In conclusion, high-quality semiconducting materials are generally grown at relatively high-temperatures, which limit the feasibility of synthetic approaches targeted to produce nanowires with controlled dimensions. We have described herein molecule-based CVD as a low-temperature approach for controlled growth of nanowires. Combining the VLS catalytic growth mode with single molecular sources allows precise control over dimensions, site-specific growth, surface states and final transport behaviors of tin oxide nanowires. Single crystalline tin oxide nanowires in array forms are synthesized in high yield at low temperatures by catalyst-assisted CVD of $[Sn(OBu^i)_4]$. The detailed structural analysis has shown the possible growth mechanism based on the VLS mechanism and crystallographic epitaxial driving force. The concept can be expanded to grow other ordered nanowires with defined growth orientation and surface facets. The achieved high quality and reproducible functional nanowires have great potential in the application of nanodevice fabrication.

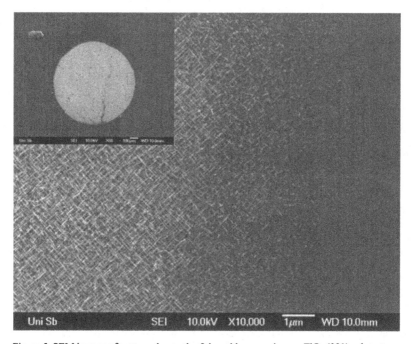

Figure 5. SEM images of patterned growth of tin oxide nanowires on TiO_2 (001) substrate.

ACKNOWLEDGEMENT

Thanks are due to the German Science Foundation (DFG) for supporting this work in the frame of the priority program on nanomaterials - Sonderforschungsbereich 277. We are thankful for the Federal Ministry of Education and Research (BMBF) for supporting this work in the frame of the priority program "BMBF-NanoFutur" (FKZ 03X5512) operating at Institute of Inorganic Chemistry, University of Cologne, Cologen, Germany.

REFERENCES

[1] YN. Xia, PD. Yang, YG. Sun, YY. Wu, B. Mayers, B. Gates, YD. Yin, F. Kim, HQ. Yan, *Adv. Mater.*, 2003, **15**, 353.

[2] JB. Hannon, S. Kodambaka, FM. Ross, RM. Tromp, *Nature,* 2006, **440**, 69.
[3] S. Mathur and H. Shen, *Encyclopedia of Nanoscience and Nanotechnology®* Ed. by H. S. Nalwa, American Scientific Publisher, 2004, **4**, 131.
[4] S. Semancik, T.B. Fryberger, *Sens. Actuators, B, Chem.,* 1990, **1**, 97.
[5] C. Tatsuyama, S. Ichimura, *Jpn. J. Appl. Phys.,* 1976, **15**, 843.
[6] W. Dazhi, W. Shulin, C. Jun, Z. Suyuan, L. Fangqing, *Phys. Rev. B,*1994, **49**, 282.
[7] Y.S. He, J.C. Campbell, R.C. Murphy, M.F. Arendt, J.S. Swinnea, *J. Mater. Res.,* 1993, **8**, 3131.
[8] Z. R. Dai, J. L. Gole, J. D. Stout, Z. L. Wang, *Adv. Funct. Mater.,* 2003, **13**, 9.
[9] S. Budak, G X. Miao, M. Ozdemir, K. B. Chetry, A. Gupta, *J. Cryst. Growth,* 2006, **291**, 405.
[10] S. Mathur, S. Barth, H. Shen, J. C. Pyun, U. Werner, *Small,* 2005, **1**, 713.
[11] Z. Liu, D. Zhang, S. Han, C. Li, T. Tang, W. Jin, X. Liu, C. Zhou, *Adv. Mater.,* 2003, **15**, 1754.
[12] A. Kolmakov, Y. Zhang, G Cheng, M. Moskovits, *Adv. Mater.,* 2003, **15**, 997.
[13] Y. J. Ma, F. Zhou, L. Lu, Z. Zhang, *Solid State Commun.,* 2004, **130**, 317.
[14] X. Jiang, Y. Wang, T. Herricks, Y. Xia, *J. Mater. Chem.,* 2004, **14**, 695.
[15] D. Li, Y. Wang, Y. Xia, *Adv. Mater.,* 2004, **16**, 361.
[16] Y. Cui, L. J. Lauhon, M. S. Gudiksen, J. Wang, and C. M. Lieber, *Appl. Phys. Lett.,* 2001, **78**, 2214.
[17] M. S. Gudiksen and C. M. Lieber, *J. Am. Chem. Soc.,* 2000, **122**, 8801.
[18] K. K. Lew and J. M. Redwing, *J. Cryst. Growth,* 2003, **254**, 14.
[19] J. Kikkawa, Y. Ohno, and S. Takeda, *Appl. Phys. Lett.,* 2005, **86**, 123109.
[20] R.S. Wagner, W.C. Ellis, *Appl. Phys. Lett.,* 1964, **4**, 89.
[21] S. G Ihn, J. I. Song, T.W. Kim, D. S. Leem, T. Lee, S. G Lee, E. K. Koh, K. Song, *Nano Lett.,* 2007, **7**, 39.
[22] A. Beltran, J. Andres, E. Longo, E. R. Leite, *Appl. Phys. Lett.,* 2003, **83**, 635.
[23] H. Wakabayashi, T. Suzuki, Y. Iwazaki, M. Fujimoto, *Jpn. J. Appl. Phys.,* 2001, **40**, 6081.

MULTIFUNCTIONAL SILICON NITRIDE CERAMIC NANOCOMPOSITES USING SINGLE-WALLED CARBON NANOTUBES

Erica L. Corral
The University of Arizona, Materials Science and Engineering Department
P.O. Box 210012
Tucson, Arizona, 85721-0012

ABSTRACT

High-temperature ceramics, such as silicon nitride, are considered the best-suited materials for use in extreme environments because they posess high melting temperatures, high strength and toughness, and good thermal shock resistance. The goal of this research is to create bulk multifunctional high-temperature ceramic nanocomposites using single-wall carbon nanotubes in order to tailor electrical and thermal conductivity properties, while also enhancing the mechanical properties of the monolith. Colloidal processing methods were used to develop aqueous single-walled carbon nanotube (SWNT)-Si_3N_4 suspensions that were directly fabricated into bulk parts using a rapid prototyping method. High-density sintered nanocomposites were produced using spark plasma sintering, at temperatures greater than 1600 °C, and evidence of SWNTs in the final sintered microstructure was observed using scanning electron microscopy and Raman spectroscopy. The multifunctional nanocomposites show exceptional fracture toughness (8.48 MPa-m$^{1/2}$) properties and was directly measured using conventional fracture toughness testing methods (ASMT C41). Our results suggest that the use of SWNTs in optimized sintered ceramic microstructures can enhance the toughness of the ceramic by at least 30% over the monolith. In addition, the observation of hallmark toughening mechanisms and enhanced damage tolerance behavior over the monolith was directly observed. The nanocomposites also measured for reductions in electrical resistivity values over the monolith, making them high-temperature electrical conductors. These novel nanocomposites systems have enhanced electrical conductivity, and enhanced toughness over the monolith which make them unique high-temperature multifunctional nanocomposites.

INTRODUCTION

Single-walled carbon nanotubes (SWNTs) are considered to be ideal reinforcement fibers for composite materials[1] because they posses exceptional mechanical (E >1TPa and TS > 7 GPa)[2] thermal (κ~1750-5800 W-m^{-1}-K^{-1})[2-4], and electrical properties (σ~10^6 S-m^{-1})[2; 5]. They also have high aspect ratios (1,000 up to 10,000), which is critical for their use in the design of advanced nanocomposite material systems. SWNTs are thermally stable in air only up to approximately 600 °C[6] and then they oxidize, which makes them difficult to process into a high-temperature ceramic matrix material where sintering temperatures are above 1400 °C. Therefore, there has been limited research effort in developing carbon nanotube (CNT)-ceramic composites[7-11] especially in high-temperature ceramic such as, Si_3N_4. The advantages of using SWNTs in a high-temperature ceramic are manifold: 1) exceptional control over electrical conductivity at temperature 2) thermal management at elevated temperatures and 3) enhanced mechanical properties of the ceramic mainly by increasing the material's resistance to brittle fracture using classical fiber reinforcing mechanisms. Ultimately, these novel ceramic nanocomposites will impact high-performance components in aerospace and other extreme temperature environmental applications. The focus of this paper will be on discussing the multifunctional properties of SWNT-Si_3N_4 nanocomposites for potential use as high-performance and high-temperature resistant (>1500 °C) composites. Our approach combines colloidal processing methods and spark plasma sintering (SPS) to make high quality composites with minimal processing defects, which allowed us to accurately measure the effects of SWNTs on thermal, electrical, and mechanical properties of the ceramic.

MATERIALS AND EXPERIMENTAL METHODS

All of the nanocomposite and monolith specimens were made from highly dispersed powders using colloidal processing methods and then densified using high temperature spark plasma sintering (SPS). The starting Si_3N_4 powder consisted of α-phase content >90% (by mass), and an average particle size of 0.96 μm that contained <10% (by mass) sintering additives (Y_2O_3, MgO and Al_2O_3). Commercially available SWNTs (Carbon Nanotechnologies Inc., Houston, TX, USA), in powder form and purified to less than 2-wt% residual metal catalyst, were used in this investigation. A cationic surfactant, cetyltrimethylammonium bromide (C_{16}TAB) (Sigma-Aldrich Corp., St. Louis, MO, USA) was used as a dispersant throughout this study. Colloidal processing methods were used to create highly dispersed nanocomposites suspensions in the green state, which remained highly dispersed in the sintered microstructure, as recently published by our group[7] and others[12]. We used SPS (Dr. Sinter SPS-1050, Sumitomo Coal Mining Co. Ltd., Tokyo, Japan) to densify our nanocomposites using a maximum pulse current of 5000 A and maximum pulse voltage of 10V. The pulse cycle was 12 ms on and 2 ms off with a heating rate of 200 °C-min[-1]. An external pressure of 25 kN was applied as the powders were heated inside a graphite die lined with graphite foil to prevent surface reactions between the powder and the inner die wall. A vacuum of 10^{-2} Torr was used and sintered under vacuum.

Density values were measured in accordance to the Archimedes method (ASTM C 830) and the rule of mixtures was used to calculate the nanocomposite densities, based on volume (density values used were 1.1 and 3.2 g cm^{-2} for SWNTs and Si_3N_4). Phase identification of the sintered materials was performed by X-Ray diffractometry (XRD) and the relative content of α/β phase in the sintered bodies was determined from the relative intensities of selected diffraction peaks[13]. Characterization of SWNT survival in the ceramic after high temperature sintering was performed using Raman spectroscopy where we measured the signature SWNT peaks. The sintered microstructure of the materials was examined by scanning electron microscopy (SEM, JEOL 6500F, JEOL USA Inc., Peabody, MA). Room temperature electrical conductivity (σ) measurements of the SWNT-Si_3N_4 nanocomposites were made using a four-point probe technique, in observation with ASTM B193-02, using a combination of volt-meters (Hewlett Packard 4339A, High Resistance Meter (Agilent 345A 81/2 Digital Multimeter).

A test method, based on the American Society of Testing and Materials (ASTM–C 1421-01b) recommended pre-cracked beam method, was devised to investigate the effect of the experimental variables on fracture toughness of the Si_3N_4 nanocomposites. The modification in the standardized test method consisted of the design of fixtures to successfully pre-crack and conduct fracture toughness testing on the smaller test specimens. Pre-cracked fracture test specimens were examined in a Hitachi S4700 field emission gun scanning electron microscope (SEM) after carbon coating.

RESULTS

Sintering and Dispersion of Si_3N_4 and SWNT Reinforced Si_3N_4 Nanocomposites using SPS

High-density nanocomposites were obtained using SPS. Table 1 shows the density values, electrical conductivity values, and fracture toughness values measured at room temperature for 0, 1, and 2-vol% SWNT-Si_3N_4 nanocomposites. Density values ranged from 87-97% theoretical density (T.D.) and showed no trend with increasing SWNT concentration, SPS temperature or SPS time at temperature. These sintering parameters need to be studied further in order to determine what effect on sintering density SWNTs have on the final sintered microstructure. **Figure 1** shows the microstructure for 2-vol% SWNT nanocomposites after fracture toughness testing. The SWNTs are dispersed on the fractured surface and also show good dispersion in the matrix after high-temperature sintering.

Table I. Sintering density, electrical and thermal properties, and fracture toughness values for Si_3N_4 and SWNT-Si_3N_4 nanocomposites measured and calculated at room temperature.

Material	SPS Temp. (°C)	SPS Time (min)	% T.D.*	Electrical Conductivity (S-m^{-1})	Fracture Toughness (MPa-m$^{1/2}$)
Si_3N_4	1600	0	96.4	3.13E-11	5.27
	1600	3	93.1	3.65E-11	8.51
1.0 vol%	1600	0	87.3	1.86E-9	4.71
	1600	3	95.4	4.76E-5	7.24
2.0 vol%	1600	0	91.0	4.76E-5	8.48
	1600	3	90.5		6.04

*Percent Theoretical Density (g-cm^{-3})

SPS provides nice control over the phase transformation from α–to–β Si_3N_4 at 1600 °C at short periods of time. For example, at 1600 °C, 0-min hold, you obtain a primarily α–Si_3N_4 microstructure and at 1600 °C, for 3-min. hold you obtain a bimodal Si_3N_4 microstructure containing both α– and β-Si_3N_4, as seen in Figure 2, respectively. The ability to control the matrix microstructure will allow us to determine how SWNTS enhance mechanical properties in an equiaxed and bimodal microstructure. **Figure 2**, shows the XRD spectra after high-temperature sintering for 2-vol% SWNT nanocomposites sintered at 1600°C for 0- and 3-min. and at 1800 °C for 0-min. hold. These plots clearly show that the phase transformation from α–to-β-Si_3N_4 takes place in the nanocomposite material and that SWNTs do not inhibit phase transformation of the matrix material. There is also no detection of carbide peaks in the XRD spectra for the nanocomposites, suggesting that the high-temperature sintering conditions do not result in reactions between SWNTs and Si_3N_4 that would form SiC at the SWNT/matrix interface. Although, these amounts may be too low concentration to detect using XRD, these high-temperature interaction warrant further study.

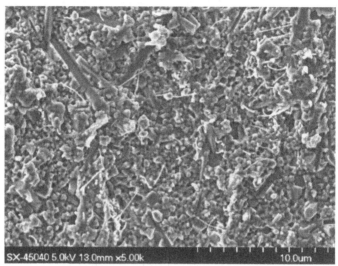

Figure 1. Scanning electron micrograph of a 2-vol% SWNT-Si₃N₄ nanocomposite after bulk fracture toughness testing (SPS: 1600 °C, 0-min. hold).

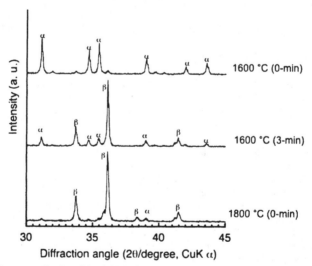

Figure 2. XRD spectra for 2-vol% SWNT-Si₃N₄ nanocomposites sintered from 1600 to 1800 °C, showing controlled phase transformation from α-to-β Si₃N₄.

SWNT Structure After High Temperature Sintering Using SPS

We used Raman spectroscopy in order to detect the signature SWNT peaks in the sintered nanocomposites and determine if any thermal decomposition of the SWNTs took place as result of high-temperature sintering. The SWNT Raman spectrum has three characteristic bands: a radial breathing mode (100-300 cm^{-1})[14; 15] a tangential mode (G-band, 1500-1600 cm^{-1})[14; 15], and a D mode[14; 15]. The defect-induced D mode originates from double-resonant Raman scattering[15] and is around 1300 cm^{-1}. However, the peak shape is indicative of metallic or semiconducting nature of the tubes and the SWNTs in this study are primarily conductive SWNTS. Our results, as seen in **Figure 3**, show that SWNTs survive high-temperature sintering up to 1600 °C, for three minutes and that the intensity of the SWNT peaks are clearly in comparison with pristine SWNT Raman signatures. Our previous work showed that SWNTs densified in a Si$_3$N$_4$ matrix can survive up to 1800 °C[7] using SPS.

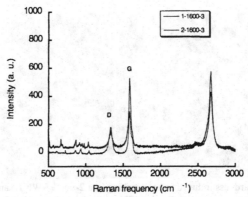

Figure 3. Raman spectroscopy shows SWNTs survive high temperature sintering using SPS for 1 and 2-vol% SWNTs sintered at 1600 °C for 3-min. hold.

Electrical Properties of SWNT Nanocomposites

SWNTs have high electrical conductivity properties that are comparable to copper ($\sigma \sim 10^6$ S-m^{-1})[2; 5]. The type of SWNT atomic bond structure can result in metallic or semiconducting properties however; most of the SWNTS used in this study were a mixture of both structure types. As should be noted, we also have SWNT bundles dispersed throughout the matrix. Experimental results show that a SWNT bundle has a longitudinal conductivity of 10^6 S-m^{-1} at 300 K[2; 16], which means we should not see an increase the electrical resistivity of the nanocomposite when used as bundles versus individual tubes. **Table 1** shows the room temperature electrical conductivity measurement values. It is well known that sintered Si$_3$N$_4$ is an excellent electrical insulator ($\sigma \sim 10^{-12}$ S-m^{-1})[17] and we found that using as little as 2-vol% SWNTs the composite went from an insulator to a semiconducting material with a measured conductivity of up to 4.76 $\times 10^{-5}$ S-m^{-1}. This value lies near the middle of the semiconducting range ($\sigma \sim 10^{-8}$–10^2 S-m^{-1}) for electrical conductivity[18].

Mechanical Behavior of SNWT-Si$_3$N$_4$ Nanocomposites

Our recent work using SWNTs as reinforcement agents in brittle ceramics has proven that further study of the bulk fracture toughness in these nanocomposite systems is warranted[19]. In this study, we used the pre-cracked beam method to measure bulk fracture toughness properties using an atomically sharp pre-crack beam specimen. We also used microhardness indentation studies to analyze the nanocomposite damage tolerance behavior over the monolith and did not use microhardness testing to calculate fracture toughness values. **Table 1** shows the fracture toughness values measured, using ASTM C-1421, for 0, 1, and 2-vol% SWNT nanocomposites. The interesting fracture toughness values are for the monolith and the 2-vol% nanocomposite specimens sintered at 1600 °C, for 0-min. where the fracture toughness increased by 30% from, 5.27 to 8.4 MPa-m$^{1/2}$, respectively. There was no measured increase in fracture toughness over the monolith using only 1-vol% SWNTs. **Figure 1** shows the fracture surface of the nanocomposites where you can see direct evidence of fiber pullout and an even dispersion of the SWNTs across the surface.

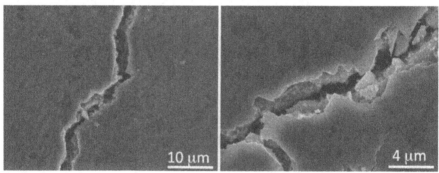

Figure 4. Microhardness induced radial cracking in a 2-vol% SWNT nanocomposites showing crack deflection, fiber pull-out and, crack bridging.

In order to measure the response of the nanocomposite to an applied load we used microhardness indentations measurement to observe the damage tolerance behavior over the monolith as a function of increasing indentation loads. The microhardness values did not deviate from nanocomposite to monolith and the average hardness values were comparable ~15 MPa. The microhardness indentations and induced cracks were analyzed for insight into mechanisms of SWNT reinforcement, such as, crack deflection, crack bridging and fiber pull-out. A load of 14N was used to induce radial cracking in the nanocomposite. At lower loads (3-10 N) there was a considerable amount of subsurface chipping and deformation around the indentation area but no radial cracking was observed. On the other hand, the monolith exhibited classic indentations with long radial cracks using applied loads much less than 15 N. Both observations suggest that ceramics with small amounts of SWNTs exhibit increased damage tolerance behavior when compared to the monolith. The SWNTs seem to impart multi-axial damage tolerance to Si$_3$N$_4$ that is novel to carbon nanotube reinforced ceramic composites. **Figure 4** shows the microhardness indentation induced radial cracks on a 2.0-vol% SWNT-Si$_3$N$_4$ composite. The three hallmark toughening mechanisms are observed in these micrographs. Crack propagation is deflected along the crack path length, fiber pull-pout is clearly seen in the wake of the crack, and crack bridging is also observed in the wake of the crack. The SWNTs are clearly stretched along the crack face and show how they are inhibiting crack propagation that leads to catastrophic failure in ceramics.

DISCUSSION

The increase in electrical conductivity may be related to the following factors. The level of dispersion achieved in the green state of the composite is retained after sintering and results in a network of SWNTs throughout the ceramic that directly enhances electrical conductivity of the ceramic. The increase in conductivity is related to a percolation of well-dispersed SWNTs at a low volume fraction, slightly above 2-vol% SWNT. When SWNTs are present in a material above the percolation limit, they also provide a continuous electrical conduction path that are not dependent on SWNT alignment or orientation within the matrix therefore, this random distribution of SWNTs in the ceramic leads to a systematic increase in electrical conductivity[20].

The fracture toughness comparison between the monolith and the 2-vol% nanocomposite, sintered at 1600 °C for 0-minutes, which have fracture toughness values of 5.27 and 8.48 MPa-$m^{1/2}$, respectively, show the potential for enhancing toughness of brittle ceramics using SWNTs. This comparison allows for the effect of SWNTs to be considered in the same sintered matrix-microstructure. Therefore, the addition of an optimized amount of SWNTs, for a particular sintering condition and sintered microstructure, clearly increases the fracture toughness of the monolithic ceramic. Although, the density values are 96 and 91% TD, for the monolith and nanocomposite, respectively, these measurements suggest that nanocomposites that have a high surface areas of reinforcement material, such as SWNTs, may not have to be sintered to full density in order to enhance mechanical properties. The overall trend observed that showed increases in fracture toughness with increases SPS time at temperature is explained by the presence of ß-Si_3N_4 grains in the matrix acting as toughening agents. It may be concluded that changing the sintering temperature and holding time cause a variation in the fracture toughness of the bulk test specimen by altering the sintered microstructure that then leads to different interactions with the SWNTs as reinforcements. Using the current processing methods, these results suggest that ß-Si_3N_4 grains have a bigger effect on fracture toughness than the SWNTs. Although, SWNTs do impart enhanced toughness in the sintered α-phase Si_3N_4 microstructure, further research is warranted in order to fully explore the role of reinforcements in sintered β-phase microstructures. After microhardness indentation induced radial cracking was observed we show direct evidence of the three hallmark toughening mechanisms used in conventional fiber reinforced ceramic matrix composites. The interaction at the interface between the SWNT and the matrix is still not clearly understood and definitely merit further study in order to understand and ultimately optimize the interfacial strength between the matrix and the fiber.

CONCLUSIONS

Multifunctional SWNT-Si_3N_4 nanocomposites were successfully made using colloidal processing and SPS methods. Our work has shown these composite have mechanically properties equal to fully sintered Si_3N_4 but that they also have the potential to increase the fracture toughness of an already structurally tough ceramic material. The insulating ceramic was transformed to be a semiconducting material at room temperature, using only very small additions of SWNTs. The Raman spectroscopy study showed that SWNTs are present in the final sintered microstructure. Also, using X-Ray diffraction analysis there was no evidence detected of any reactions between the fiber and matrix material. SPS also allows for processing high-density nanocomposite with precise control over the phase transformation of the matrix material, which has provided insight into how SWNTs are more effective in enhancing toughness in an equiaxed matrix microstructure over a rod-like or long grain microstructure. Further studies are needed in order to understand the effects of high-temperature and pressure on the structures of SWNTs in a high-temperature ceramic but the initial findings reported here suggest that there is significant potential for these novel nanocomposites in high performance applications, at temperature, with increased toughness, increases electrical conductivity and reduced thermal conductivity.

ACKNOWLEDGEMENTS

Author is grateful to the High Temperature Materials Laboratory User Program at Oak Ridge National Laboratory in Oak Ridge, Tennessee for technical assistance in nanocomposite property measurements.

REFERENCES

1. Paul Calvert, "Nanotube Composites: A Recipe for Strength," *Nature,* **399** 210-11 (1999).
2. M. Meyyappan, "Carbon Nanotubes: Science and Applications,"). Boca Raton: CRC Press LLC. (2005).
3. J. Hone, C. Piskoti, and A. Zettl, "Thermal Conductivity of Single-Walled Carbon Nanotubes," *Physics Review B,* **59** [4] R2514-R16 (1999).
4. S. Berber, Y.-K. Kwon, and D. Tomanek, "Unusually High Thermal Conductivity of Carbon Nanotubes," *Physical Review Letters,* **84** [20] 4613-16 (2000).
5. J. Hone, M.C. Llaguno, N.M. Nemes, A. T. Johnson, J.E. Fischer, D.A. Walters, M.J. Cassavant, J. Schmidt, and R. E. Smalley, "Electrical and Thermal Properties of Magnetically Aligned Single Wall Carbon Nanotube Films," *Applied Physics Letters,* **77** [5] 666-68 (2000).
6. M. Meyyappan, "Carbon Nanotubes: Science and Application," pp. 289. In. CRC Press, New York, 2005.
7. E.L. Corral, J. Cesarano, A. Shyam, E. Lara-Curzio, N. Bell, J. Stuecker, N. Perry, M. DiPrima, Z. Munir, J. Garay, and E.V. Barrera, "Engineered Nanostructures for Multifuntional Single-Walled Carbon Nanotube Reinforced Silicon Nitride Nanocomposites," *Journal of the American Ceramic Society,* **91** [10] 3129-37 (2008).
8. L. L. Yowell, "Thermal Management in Ceramics: Synthesis and Characterization of Zirconia-Carbon Nanotube Composite," pp. 1-161. In *Department of Mechanical Engineering and Materials Science.* Rice University, Houston, 2001.
9. J. Tatami, T. Katashima, K. Komeya, T. Meguro, and T. Wakihara, "Electrically Conductive Cnt-Dispersed Silicon Nitride Ceramics," *Journal of American Ceramic Society,* **88** [10] 2889-93 (2005).
10. R. Z. Ma, J. Wu, B. Q. Wei, J. Liang, and and D. H. Wu, "Processing and Properties of Carbon Nanotubes-Nano-Sic Ceramic," *Journal of Mateirals Science,* **33** 5243-46 (1998).
11. A. Peigney, Ch. Laurent, E. Flahaut, and A. Rousset, "Carbon Nanotubes in Novel Ceramic Matrix Nanocomposites," *Ceramics International,* **26** 677-83 (2000).
12. R. Poyato, A. L. Vasiliev, N. P. Padture, H. Tanaka, and T. Nishimura, "Aqueous Colloidal Processing of Single-Wall Carbon Nanotubes and Their Composites with Ceramics," *Nanotechnology,* **17** 1770-77 (2006).

13. C. P. Gazzara and D. R. Messier, "Determination of Phase Content of Si_3n_4 by X-Ray Diffraction Analysis," *American Ceramic Society Bulletin*, **56** [9] 777-80 (1977).
14. A.M. Rao, E. Richter, S. Bandow, B. Chase, P.C. Eklund, K. W. Williams, S. Fang, K.R. Subbaswamy, M. Menon, A.Thess, R.E. Smalley, G. Dresselhaus, and M.S. Dresselhaus, "Diameter-Selective Raman Scattering from Vibrational Modes in Carbon Nanotubes," *Science*, **275** 187-91 (1998).
15. M.S. Dresselhaus, G. Dresselhaus, A. Jorio, A.G. Filho Souza, M.A. Pimenta, and R. Saito, "Single Nanotube Raman Spectroscopy," *Accounts of Chemical Research*, **35** [12] 1070-78 (2002).
16. A. Thess, R. Lee, P. Nikolaev, H. Dai, P. Petit, J. Robert, C. Xu, Y. H. Lee, S. G. Kim, G. Rinzler, D. T. Colbert, G. Scuseria, D. Tomanek, J. E. Fischer, and R. E. Smalley, "Crystalline Ropes of Metallic Carbon Nanotubes," *Science*, **273** 483-87 (1996).
17. Frank L. Riley, "Silicon Nitride and Related Materials," *Journal of the American Ceramic Society*, **83** [2] 245-65 (2000).
18. Rolf E. Hummel, Electronic Properties of Materials. New York: Springer-Verlag. (2000).
19. A. Peigney, "Composite Materials: Tougher Ceramics with Nanotubes," *Nature Materials*, **2** [January 2003] 15-16 (2003).
20. G.-D. Zhan, J. D. Kuntz, J. E. Garay, and A. K. Mukherjee, "Electrical Properties of Nanoceramics Reinforced with Ropes of Single-Walled Carbon Nanotubes," *Applied Physics Letters*, **83** [6] 1228-30 (2003).
21. Guo-Dong Zhan, Joshua D. Kuntz, Hsin Wang, Chong-Min Wang, and Amiya K. Mukherjee, "Anisotropic Thermal Properties of Single-Wall-Carbon-Nanotube Reinforced Nanoceramics," *Philosophical Magazine Letters*, **84** [7] 419-23 (2004).
22. G.-D. Zhan, J. D. Kuntz, J. Wan, and A. K. Mukherjee, "Single-Wall Carbon Nanotubes as Attractive Toughening Agents in Alumina Based Nanocomposites," *Nature Materials*, **2** 38-42 (2003).
23. G.-D. Zhan, J. D. Kuntz, H. Wang, C.-M. Wang, and A. K. Mukherjee, "Anisotropic Thermal Properties of Single-Wall-Carbon-Nanotube-Reinforced Nanoceramics," *Philosophical Magazine Letters*, **84** [7] 419-23 (2004).
24. S. Torquato and M.D. Rintoul, "Effect of the Interface on the Properties of Composite Media," *Physics Review Letters*, **75** [22] 4067-70 (1995).
25. J. P. Issi, "Graphite and Precursors,"). New York: Gordon and Breech. (2001).

SIMULATION BASED DESIGN OF POLYMER CLAY NANOCOMPOSITES USING MULTISCALE MODELING: AN OVERVIEW

Dinesh R. Katti and Kalpana S. Katti

Department of Civil Engineering, North Dakota State University
Fargo, North Dakota, U.S.A

ABSTRACT

This paper provides an overview of our work on polymer clay nanocomposites that has led to a new theory for mechanical property enhancement in these nanocomposites. We refer to this theory as the 'altered phase theory'. A systematic approach that combines experimental and modeling techniques at various length scales to probe and understand mechanisms leading to property enhancement is reported. This study has provided an insight into the key role of molecular interactions on property enhancement in nanocomposites. Results of our simulations spanning in scale from molecular dynamics to finite element method provide quantitative foundation for the new theory. This work also describes new modeling techniques for MD simulations to represent real nanocomposite systems and innovative use of atomic force microscopy to evaluate altered phases in nanocomposite. Interaction energy maps of molecular interactions between various constituents of nanocomposites provide an excellent view into how molecular interactions occur in nanocomposites and describe their role in macroscopic properties. Here, we present a summary of our work of which details can be found in various papers published elsewhere and referred to in this paper. Figures in this paper are from our published work and are appropriately referenced.

INTRODUCTION

Polymer clay nanocomposites are materials of significant interest because of their potential to cause dramatic increase in mechanical and thermal properties over the properties of pristine polymer, by addition of small amounts of nanosized clay particles in the polymer. However, specific mechanisms leading to these improvements in properties by addition of small amounts of nanosized clays are not understood. Thus, designing such materials with tailored properties is currently out of reach. Our research has focused on evaluating the specific mechanisms responsible for property enhancement in polymer clay nanocomposites. Clays are hydrophilic in nature and many polymers of interest in structural applications are hydrophobic. Organic modifiers are used to make the clays compatible and intercalate or exfoliate the clay. In the work, we describe a single clay system (montmorillonite), one polymer (nylon6) and three organic modifiers. This work provides an overview of the work done by our group that has resulted in a new 'altered phase theory' for polymer clay nanocomposites.

NATURE OF MOLECULAR INTERACTIONS IN PCN[1]

Fourier transform infrared spectroscopic studies were conducted on organically modified sodium montmorillonite and polymer clay nanocomposite samples to evaluate the nature of molecular interactions between clay, organic modifiers and polymer. The montmorillonite used in this study was swy-2, crook county, Wyoming obtained from the clay minerals repository. Organic modifiers used were 12-aminolauric acid (aminol / lauric), n-dodecylamine (dodecyl) and 1,12-diaminododecane (dodecane). The polymer used in the study was Polyamide6 (PA6) $[(CH_2)_5CONH]_n$ of molecular weight 16,000.

Some of the key conclusions from the FTIR studies are 1) adsorbed water in the interlayer is replaced by organic modifier, 2) formation of hydrogen bonding between functional group of modifier and surface oxygen of silica tetrahedra, 3) shift in Si-O in-plane stretching band of MMT indicating non-bonded interactions between clay and the modifier in OMMT as well as PCN, 4) formation of hydrogen bond between amide group of PA6 and the functional group of the modifier and 5) no new bonded interactions were observed either in OMMT or the PCN.

MD MODEL DEVELOPMENT[2]

A new method was developed to construct realistic models for MD simulation of OMMT and PCN. Most MD models are based purely on minimum energy consideration. Here we use an innovative approach wherein we use a combination of experimental results from X-ray diffraction, MD simulations and FTIR experiments to construct models that are representative of real OMMT and PCN. Various starting d-spacing were used and the models were minimized. The model with closest d-spacing to that obtained from X-ray diffraction experiments and corresponding lowest minimized energy was chosen as the representative model. This was done for all modifiers. This methodology allows for more accurate interpretation of results from MD simulations for real systems. This study also provided information about the size of polymer intercalating in the clay gallery. The parameters for the clay model were found in our previous study.[3-5]

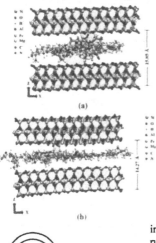

Figure 1. (a) Initial PCN model (b) Final PCN model after MD simulation. The organic modifiers are in ball and stick model, and the polymer is in licorice rendering form. [2]

MAPS OF INTERACTION ENERGIES BETWEEN CONSTITUENTS[6]

Using the energy evaluation tool, MDEnergyTM of NAMD 2.5, the interaction energies between different constituents of composites are found. The trajectory file of the whole molecular model is directly obtained from MD simulation. The interactions energies for any molecular system can be calculated for bonded and nonbonded energies separately. Further, the bonded energies can be computed in the category of bond, angle, and dihedral energies in the molecular system. Similarly, the nonbonded energy can be computed by splitting into van der Waals, and electrostatic energies for any given set of atoms or between two sets of interacting atoms. All MD simulations of OMMT and PCNs are run for duration of 200 pico second (ps) in the final stage to equilibrate the models. The average of results for last 25 ps was considered for calculating the bonded and different non-bonded energies of the molecular models.

Using the representative molecular models developed above, interaction energies between the three constituents of the PCN were evaluated using molecular dynamics software NAMD. The interaction energies are nonbonded in nature. For each of the interaction pairs, both Van der Walls and electrostatic components of interaction energies were calculated. Also, for the modifier and the polymer, contributions from functional groups and backbone chains were also evaluated.

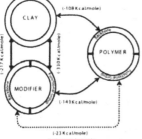

Figure 2. Schematic representation of significant attractive non-bonded interactions in PCN. [6]

Interaction energy maps were drawn based on the results (Figure 2). Our results show very strong attractive interactions between the modifier and the clay. The major interaction of the polymer with the modifier is via the functional groups. Major interaction between the polymer and clay occurs through the backbone of the polymer as a result of collective effect of large number of backbone chain hydrogen atoms.

INFLUENCE OF ORGANIC MODIFIERS ON CRYSTALLINITY AND NANOMECHANICAL PROPERTIES[7]

Experimental studies were conducted to evaluate crystallinity of PCN containing the same (9%wt.) clay, the same polymer but different organic modifiers using differential scanning calorimetry (DSC). For comparison, the pure polymer was subject to same synthesis route as adopted for synthesis of PCN. Results reveal that the percent crystallinity differs for the three PCNs and the polymer. The percent crystallinity of pure polymer being the highest. Nanomechanical static and dynamic tests were conducted on these samples. Results show a clear relationship between nanomechanical properties and crystallinity. We have evaluated the dynamic mechanical properties of PCNs in the nanometer length scale using nano dynamic mechanical analyzer (NanoDMA). The elastic modulus and hardness of PCNs in the nanometer length scale have been evaluated using a nanoindenter. The elastic modulus is highest for PCN with lowest crystallinity and decreases with increasing crystallinity with the lowest elastic modulus for the pure polymer (Figure 3). Similar relationships were observed for storage and loss modulus values obtained from dynamic nanoindentation tests. Samples for the same organic modifier and polymer but with different amounts of clay showed that with increasing clay loading, the crystallinity of PCN decreases. As before, the nanomechanical properties increase with decreasing crystallinity of the PCN. These results provide a qualitative indication that molecular interactions play an important role in crystallinity and nanomechanical properties of PCN.

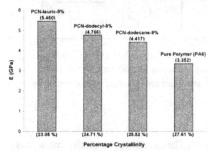

Figure 3. Relationship between nanomechanical elastic modulus and crystallinity in PCN. Different organic modifiers alter crystallinity of PCN by different amounts. [7]

MOLECULAR INTERACTIONS INFLUENCE CRYSTALLINITY AND NANOMECHANICAL PROPERTIES[8]

Molecular dynamics simulations were conducted on three PCN models containing three different organic modifiers. These models represent the three PCN samples described and tested in the previous section. Interaction energies between constituents were calculated. Interaction energies between the constituents were different for the three PCNs. For the case of PCN-lauric which shows the least amount of crystallinity and highest nanomechanical properties, the attractive interaction energy between polymer functional group and modifier functional group is the highest while at the same time, the repulsive interaction energy between modifier and backbone of the polymer is the highest (Figure 4). For the case of PCN-dodecane where crystallinity is maximum among the three PCNs, and the nanomechanical properties are the smallest, the attractive interaction energy between polymer functional group and modifier functional group is the least, and at the same time the repulsive interaction energy between the modifier and the polymer backbone chain is the least. The simultaneous attractive interaction at the functional group of the polymer and repulsive interaction with the backbone chain of the polymer results in disturbance to the periodicity of the polymer structure, causing reduction in crystallinity. The high attractive interactions between functional group of modifier and polymer also contribute to enhancement of mechanical properties. Figure 5 shows a schematic diagram of the concept.

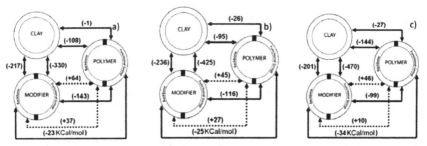

Figure 4. Interaction energy maps of PCN with three different organic modifiers a) PCN-lauric, b) PCN-dodecyl and c) PCN-dodecane. The negative values are attractive interaction energies and the positive values are repulsive interaction energies. [8]

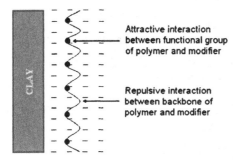

Attractive interaction between functional group of polymer and modifier

Repulsive interaction between backbone of polymer and modifier

Figure 5. The simultaneous attractive interactions between functional group of polymer and modifier and repulsive interaction between backbone of polymer and modifier results in disturbing the periodicity of polymer and thus changing its crystallinity. [8]

ALTERED PHASE MODEL FOR PCN[9, 10]

Photoacoustic spectroscopy of the three organically modified clays was performed to study the influence of organic modifiers and their interactions with clay to the structure of clay. Specifically, the Si-O region of the spectrum was observed to be influenced significantly by the modifiers[9]. The three modifiers influence the Si-O regions differently indicating that the molecular interactions seem to influence the clay structure.

Thus it appears that the molecular interactions between the constituents result in altering both phases of the nanocomposite, the polymer phase and the clay. We propose a new altered phase theory for polymer clay nanocomposites that would explain the reasons for property enhancement in PCN. Figure 6 is a schematic diagram showing the concept.

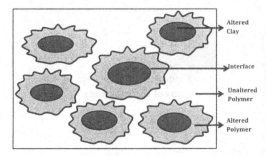

Figure 6. Schematic of altered phase model for PCN. [10]

PROPERTIES AND EXTENT OF ALTERED POLYMER IN PCN USING MULTISCALE TECHNIQUES. [10]

A multiscale experimental and computational approach was used to evaluate the physical extent of the altered zone and find elastic modulus of the altered zone. Experiments involved atomic force microscopy and modeling involved steered molecular dynamics and finite element method. In steered molecular dynamics, force or velocity can be applied to individual atom or atoms in MD and resulting response is obtained to evaluate properties. Steered molecular dynamics simulations were used to evaluate the elastic properties of clay sheets, interaction zone and intercalated polymer and modifier (Figure 7). The properties obtained were introduced into a finite element model of a block of PCN. Two simulation conditions were studied 1) considering PCN as a conventional composite model and 2) incorporating the altered phase theory in the model. In both cases, a small normal stress was applied to the PCN finite element block model to evaluate the elastic modulus of PCN. The conventional composite model failed to predict the large increase in elastic modulus obtained experimentally. The simulation showed only a small increase relative to elastic modulus of the pure polymer.

In order to incorporate the altered phase model, the extent of the altered zone was found using phase imaging from atomic force microscopy test (Figure 8). The dark brown regions of the image indicate polymer phase, and the light regions indicate the clay. The altered phase is the lighter brown region between the clay sheets and the polymer region. The average thickness of this region was found to be 25 nm. The altered phase model was then incorporated into the finite element model 9Figure 9)

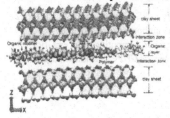

and a range of values of elastic modulus of the altered zone were introduced and the elastic modulus of the PCN was computed. The elastic modulus of the altered zone was found to be 20 GPa as against 3.35 GPa for the polymer.

Figure 7. Molecular model of intercalated PCN for conducting steered molecular dynamics simulations to evaluate nanoscale mechanical properties. [10]

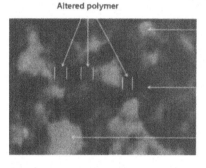

Figure 8. AFM phase imaging showing extent of altered polymer. [10]

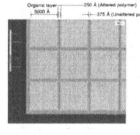

Figure 9. Cross-section through the finite element model showing clay, altered polymer and unaltered polymer zones. [10]

CONCLUSIONS

A multiscale experimental and computational approach was used to find the mechanisms responsible for enhancement of mechanical properties in polymer clay nanocomposites. Molecular interactions between constituents in nanocomposite systems alter phases in the system in a dramatic way to a very significant distance. Thus small amounts of well dispersed nanoparticles can alter properties of significant volume of the matrix resulting in large increase in mechanical properties observed in such systems. In nanocomposite systems, molecular interactions are instrumental to property enhancement. We have also observed significant influence of nonbonded interactions in biological nanocomposite systems such as nacre[11-13] and bone[14-16]

ACKNOWLEDGEMENTS

The authors would like to thank National Science Foundation (NSF) grants 0556020 , 0114622, 0320657 and the NSF program manager Dr. Richard Fragaszy. The authors acknowledge the computational hardware support by NDSU-Center for high performance computing and TeraGrid resource allocation grant. Hardware support by Dr. Gregory Wettstein of CHPC is acknowledged.

REFERENCES

1.   Katti, K. S.; Sikdar, D.; Katti, D. R.; Ghosh, P.; Verma, D., Molecular interactions in intercalated organically modified clay and clay-polycaprolactam nanocomposites: Experiments and modeling. *Polymer* **2006,** *47* (1), 403-414.
2.   Sikdar, D.; Katti, D. R.; Katti, K. S., A molecular model for epsilon-caprolactam-based intercalated polymer clay nanocomposite: Integrating modeling and experiments. *Langmuir* **2006,** *22* (18), 7738-7747.
3.   Katti, D. R.; Schmidt, S. R.; Ghosh, P.; Katti, K. S., Modeling the response of pyrophyllite interlayer to applied stress using steered molecular dynamics. *Clays and Clay Minerals* **2005,** *53* (2), 171-178.
4.   Schmidt, S. R.; Katti, D. R.; Ghosh, P.; Katti, K. S., Evolution of mechanical response of sodium montmorillonite interlayer with increasing hydration by molecular dynamics. *Langmuir* **2005,** *21* (17), 8069-8076.
5.   Katti, D. R.; Schmidt, S. R.; Ghosh, P.; Katti, K. S., Molecular modeling of the mechanical behavior and interactions in dry and slightly hydrated sodium montmorillonite interlayer. *Canadian Geotechnical Journal* **2007,** *44* (4), 425-435.
6.   Sikdar, D.; Katti, D. R.; Katti, K. S.; Bhowmik, R., Insight into molecular interactions between constituents in polymer clay nanocomposites. *Polymer* **2006,** *47* (14), 5196-5205.
7.   Sikdar, D.; Katti, D.; Katti, K.; Mohanty, B., Effect of organic modifiers on dynamic and static nanomechanical properties and crystallinity of intercalated clay-polycaprolactam nanocomposites. *Journal of Applied Polymer Science* **2007,** *105* (2), 790-802.
8.   Sikdar, D.; Katti, D. R.; Katti, K. S., The role of interfacial interactions on the crystallinity and nanomechanical properties of clay-polymer nanocomposites: A molecular dynamics study. *Journal of Applied Polymer Science* **2008,** *107* (5), 3137-3148.
9.   Sikdar, D.; Katti, K. S.; Katti, D. R., Molecular interactions alter clay and polymer structure in polymer clay nanocomposites. *Journal of Nanoscience and Nanotechnology* **2008,** *8* (4), 1638-1657.
10.   Sikdar, D.; Pradhan, S. M.; Katti, D. R.; Katti, K. S.; Mohanty, B., Altered phase model for polymer clay nanocomposites. *Langmuir* **2008,** *24* (10), 5599-5607.
11.   Katti, K. S.; Katti, D. R.; Pradhan, S. M.; Bhosle, A., Platelet interlocks are the key to toughness and strength in nacre. *Journal of Materials Research* **2005,** *20* (5), 1097-1100.
12.   Ghosh, P.; Katti, D. R.; Katti, K. S., Mineral proximity influences mechanical response of proteins in biological mineral-protein hybrid systems. *Biomacromolecules* **2007,** *8* (3), 851-856.
13.   Ghosh, P.; Katti, D. R.; Katti, K. S., Mineral and protein-bound water and latching action control mechanical behavior at protein-mineral interfaces in biological nanocomposites. *Journal of Nanomaterials* **2008**.
14.   Bhowmik, R.; Katti, K. S.; Katti, D. R., Mechanics of molecular collagen is influenced by hydroxyapatite in natural bone. *Journal of Materials Science* **2007,** *42* (21), 8795-8803.
15.   Bhowmik, R.; Katti, K. S.; Verma, D.; Katti, D. R., Probing molecular interactions in bone biomaterials: Through molecular dynamics and Fourier transform infrared spectroscopy. *Materials Science & Engineering C-Biomimetic and Supramolecular Systems* **2007,** *27* (3), 352-371.
16.   Bhowmik, R.; Katti, K. S.; Katti, D. R., Influence of mineral on the load deformation behavior of polymer in hydroxyapatite-polyacrylic acid nanocomposite biomaterials: A steered molecular dynamics study. *Journal of Nanoscience and Nanotechnology* **2008,** *8* (4), 2075-2084.

PREPARATION AND CHARACTERISTIC CONTROL OF CONDUCTING POLYMER/METAL OXIDE NANO-HYBRID FILMS FOR SOLAR ENERGY CONVERSION

Yasuhiro Tachibana,* Satoshi Makuta, Yasuhide Otsuka, Jun Terao, Susumu Tsuda, Nobuaki Kambe and Susumu Kuwabata

Department of Applied Chemistry, Graduate School of Engineering, Osaka University
2-1 Yamada-oka, Suita, Osaka 565-0871, Japan
Tel: +81-6-6879-7374, Fax: +81-6-6879-7374, e-mail: y.tachibana@chem.eng.osaka-u.ac.jp

ABSTRACT

We review our recent investigation of photoinduced polymerization of thiophene inside TiO_2 nanopores to form nanostructured polythiophene/TiO_2 heterojunction films and the related kinetic studies. The nanohybrid film possesses dense heterojunction and electronic connection within the TiO_2 nanoporous domain. Photo-polymerization proceeded in 3 stages, (i) photoexcitation of bithiophene covalently attached to the TiO_2 surface, (ii) an electron injection reaction from the surface attached thiophene to the TiO_2 and (iii) an electron transfer from a thiophene reactant in an electrolyte to the surface attached oxidized bithiophene. The nanohybrid film was applied to a sensitized-type photoelectrochemical solar cell, substantiating direct application of the nanohybrid film to electronic devices. Despite increase in light harvesting efficiency, wavelength dependent incident photon-to-current conversion efficiency decreased with the light irradiated polymerization time. In order to identify a factor controlling photocurrent efficiency, kinetic studies at TiO_2/bithiophene/electrolyte interfaces were conducted, and their parameters to the solar cell functions were related. Comparison of emission studies between the bithiophene adsorbed TiO_2 and Al_2O_3 revealed the electron injection from the excited bithiophene into the TiO_2 with the efficiency of nearly 100 %. The charge recombination between the bithiophene cation and the electron in the TiO_2 appeared to be fast with a half decay time of 70 µs in comparison to the ruthenium dye sensitized TiO_2 film (~1 ms). The bithiophene regeneration kinetics was slightly faster, clarifying the inferior photocurrent performance.

INTRODUCTION

Organic/Inorganic polymer hybrid materials have a wide variety of attractive potentialities to introduce novel structural design in material sciences[1-3] and to derive novel functions for device applications.[4-6] In particular, a hybrid structure based on a conducting polymer/metal oxide semiconductor is one of the most advantageous combinations for photo-electronic devices. As organic materials, conducting polymers possess distinctive properties such as economical viability, light weight and easy processability, and high suitability for various types of electronic devices, e.g. display devices,[7] lasers,[8] FETs,[9] and photovoltaics.[10] In contrast, metal oxides such as titanium dioxide, zinc oxide and tin oxide are environmentally viable with excellent chemical and physical stability, having been studied for transparent electrodes,[11] electrochromism,[12] photocatalysis[13] and solar cells.[14] By combination of these materials with nanometer size control, an efficient electron transfer reaction or energy transfer reaction can be derived,[6, 15-17] i.e. introducing further novel functions in addition to their individual attractive properties.

Application of conducting polymer/metal oxide nanometer size controlled hybrids, nanohybrids, to photovoltaic devices has recently been studied. Bulk heterojunction type is one of the most attractive structures since enhanced light absorption, charge separation and charge transportation can be achieved.[16, 18-24] Sensitization type photoelectrochemical cells using a liquid electrolyte have also been developed owing to their simple fabrication process.[25-29]

Fabrication of polymer/metal oxide nanohybrid films has generally been accomplished using the blend method and the penetration method. In the blend method, conducting polymer and metal oxide nanoparticles are mixed in solution, and subsequently nanostructured films were prepared by spin coating.[16, 21, 22] This method is attractive owing to the simple process, and can be introduced to diversified applications. In addition, dense hybridization throughout the film can be expected. For the penetration method, metal oxide nanostructured films were first prepared, and then the conducting polymer solution was placed on top of the metal oxide film. The conducting polymer penetrates into the pores of the film until the polymer solution evaporates.[17, 19, 23, 24] In this method, the electronic connection within the metal oxide domain is maintained with the hybridization occurring with ease, facilitating the device assembly.

For the application of the polymer/metal oxide nanohybrid films to electronic devices, a dense heterojunction (hybridization) and an electronic connection within the metal oxide throughout the film are essential, since the device function is often influenced by charge transfer reaction at the polymer/metal oxide interface and by charge carrier transport through the metal oxide domain. In this regard, using the blend method, one may find difficulty to obtain the electric contact between the metal oxide particles throughout the film, since the polymer chain may interfere with the connection formed within the metal oxide. Complete hybridization may not be obtained in the penetration method, as the polymer may not be filled owing to limited penetration depth of the conducting polymer layer. For example, Huisman et al. reported the penetration depth of 1 μm when dodecylthiophene is employed to fill inside the pores in the TiO_2 nanocrystalline film.[19]

As a novel fabrication process of polymer/metal oxide nanohybrid films, we have recently introduced photoinduced polymerization of thiophene inside the TiO_2 nanoporous film.[30] Initially, thiophene monomers possessing a carboxyl group, 2,2'-bithiophene-5-carboxylic acid (BTC), were adsorbed on the TiO_2 surface, and a selective excitation of the BTC led to the thiophene photo-polymerization inside the TiO_2 pore. Thus, this method has thoroughly provided nanohybridization and electronic connection within TiO_2. The resultant nanohybrid films were readily applied as working electrodes in sensitized-type solar cells. This experimentation was devised to substantiate if this nanohybrid fabrication method can directly be employed for device applications.

In this paper, we briefly review photo-induced fabrication processes of polythiophene/TiO_2 nanohybrid films and the detailed relationship between interfacial kinetics and performance of the solar cells based on the nanohybrid film. For the kinetic studies, we particularly focused on TiO_2/BTC/electrolyte interfaces.

EXPERIMENTAL

Samples

TiO$_2$ nanocrystalline films, thickness 4~7 μm, were prepared on a slide glass or a fluorine doped tin oxide glass, FTO, (Asahi glass, type-U, 10 Ω/square) as previously described.[30] The TiO$_2$ paste (Ti-Nanoxide HT/SP, 15 nm in diameter) was purchased from Solaronix SA. The film, after printing, was leveled for 15 min, heated up to 500 ℃ at 15.8 ℃/min, and calcined at 500 ℃ for 1 h in an air flow oven. Al$_2$O$_3$ (Alu C, γ-Al$_2$O$_3$: 67%, δ-Al$_2$O$_3$: 33%, tetragonal crystal system, particle diameter of approximately 13 nm) nanoporous films, thickness: 4~7 μm, were prepared following the reported method.[30] The printed Al$_2$O$_3$ films were calcined at 500 ℃ for 1 h in air. Sensitization of a TiO$_2$ or Al$_2$O$_3$ film by 2,2'-bithiophene-5- carboxylic acid, BTC, (Maybridge, 97%) was conducted by immersion into the 10 mM BTC ethanol solution for 10 min. at room temperature.

Photo-polymerization and electrochemical measurements were performed in 1-butyl-3-methylimidazolium bis(trifluoromethanesulfonyl)imide, BMITFSI, an ionic liquid, or in 0.1 M lithium perchlorate, LiClO$_4$, (Wako) in propylene carbonate, PC, (Wako). BMITFSI was synthesized following the previously reported method,[31, 32] and dried at 105 ℃ in vacuum for at least 1 h.

Photo-polymerization of thiophene

Polythiophene was photoelectrochemically synthesized from BTC adsorbed on TiO$_2$ (BTC/TiO$_2$). Photo-polymerization was performed by exciting the BTC, after the BTC/TiO$_2$ electrode was placed in a three-electrode cell, containing a thiophene reactant in the electrolyte, under white light from a Xe lamp through a >400 nm high wavelength pass filter (320 mW/cm^2) with application of +0.5 V vs. Ag/AgCl. This filter was intentionally inserted in the light path to avoid the TiO$_2$ excitation by UV light. A Pt plate (0.4 cm^2) and an Ag/AgCl electrode were used as a counter electrode and a reference electrode, respectively. After the photo-polymerization, the hybrid film was washed with ethanol and dried in air.

Characterization of nanohybrid films

Chronoamperometry using a potentiostat (Hokuto Denko, HSV-100) were employed to detect photocurrents from the TiO$_2$ electrode. The amount of electrons collected from the TiO$_2$ electrode was calculated by integrating the data to estimate the photo-polymerization yield. Absorption and emission spectra were measured by a UV/Vis absorption spectrometer (JASCO, V-670) and an emission spectrometer (Horiba, FluoroLog-3), respectively. Morphology of the nanohybrid film was observed by FE-SEM (Hitachi, S-5200).

The oxidation potential of the BTC was determined as the onset potential of the thiophene oxidation in linear sweep voltammograms (ALS 660C) using a three-electrode cell consisting of a gold working electrode (diameter: 100 μm), a Pt plate counter electrode and an Ag/AgCl reference electrode in PC with 0.1M LiClO$_4$. All electrochemical data were presented against the Ag/AgCl potential.

Optical experiments

Microsecond to millisecond transient absorption studies were conducted with a Nd/YAG laser

(Spectra Physics, Quanta-Ray GCR-11) pumped dye laser (Usho Optical Systems, DL-100, ~10 ns pulse duration) as a pump source, a 100 W tungsten lamp as a probe source, and a photodiode-based detection system (Costronics Electronics) with a TDS-2022 Tektronix oscilloscope. Details of the laser system will be described elsewhere. The BTC sensitized TiO_2 (BTC/TiO_2) film was immersed in an optical cuvette filled with approximately 4 ml propylene carbonate solution containing 0.1 M LiI or $LiClO_4$. The Li^+ concentration was adjusted to 0.1 M, so that the TiO_2 conduction band energy level remained constant.[33-35] Transient data were obtained by employing a low excitation density ~0.13 mJ/cm^2 with 425 nm excitation at 1 Hz. This excitation density corresponds to ~1.0 excited BTC per TiO_2 nanoparticle.

Photoelectrochemical solar cells

Sandwich type solar cells were fabricated by binding a redox electrolyte with the nanohybrid electrode and the Pt counter electrode.[36] Prior to the BTC adsorption on the TiO_2, the film was treated with titanium tetrachloride aqueous solution to coat the surface with a thin TiO_2 layer. The electrolyte was prepared by dissolving 0.6 M dimethyl propyl imidazolium iodide (Tomiyama Pure Chemical), 0.05 M iodine (99.8%, Wako), 0.1 M lithium iodide (99.995 %, Wako) and 0.5 M *tert*-butylpyridine (99%, Aldrich) in dried acetonitrile (99%, Wako). A Xe lamp with a monochromator (Bunko Keiki, SM-25) was used for IPCE measurements. J-V curves were observed by an AM1.5 solar simulator (100 mW/cm^2, Yamashita Denso, YSS-50A).

RESULTS AND DISCUSSION

Formation of polythiophene/TiO_2 nanohybrid films

Prior to photoelectrochemical polythiophene formation, attachment of a visible light sensitive thiophene molecule on the TiO_2 was investigated to achieve a heterojunction between a polymerized thiophene and the TiO_2. A molecule possessing a carboxyl group is known to form a chemical bond with the TiO_2 surface,[14] and thus we considered thiophene molecules with a carboxyl group. We found by introducing a commercially available BTC on the TiO_2 surface, the film turned yellow. Figure 1 compares an absorption spectrum and a photograph of the BTC/TiO_2 with those of a TiO_2 film alone. The yellow coloration may originate from a BTC-TiO_2 charge transfer (CT) state (from BTC HOMO to Ti^{4+}) following the tight adsorption of the BTC molecules to the TiO_2 surface. Appearance of similar absorption bands were previously reported,[37, 38] and were assigned to CT absorption bands. An absorption spectrum of 100 μM BTC in ethanol exhibits no coloration with light absorption at <380 nm, as shown in Figure 1.

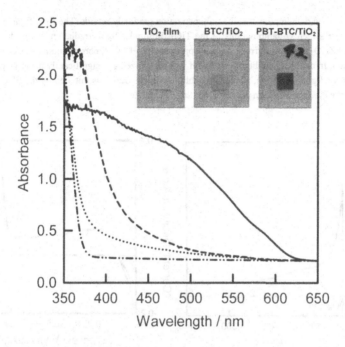

Figure 1. Absorption spectra and photographs (inset) of the TiO_2 film (·······), BTC/TiO_2, (– – –) and PBT-BTC/TiO_2 (——). The paste containing large TiO_2 particles (diameter: ~50 nm) was used for the photographs. An absorption spectrum of 100 μM BTC in ethanol (–··–··–) is also shown as comparison.

Photo-polymerization was performed by exposing light (>400 nm) to the cell in order to excite only the surface attached BTC with application of +0.5 V. As the light continuously irradiated, the color of the film, in contact with the electrolyte, turns from yellow to dark blue, typically observed for doped polythiophene.[39] Note that the bias application in the range of 0~+0.5 V is necessary to collect the electrons for the oxidation polymerization. This dark blue color readily turned red by reducing the doped thiophene with application of a negative bias for a long time. An absorption spectrum of a dedoped polythiophene/TiO_2 nanohybrid film, PBT-BT/TiO_2, prepared with 0.1M bithiophene as a thiophene reactant in BMITFSI is shown in Figure 1. In this case, the polymerization was performed for 80 min. The absorption is featureless and broad at <620 nm, indicating that the film possesses a wide range of polymerization degree, i.e. mixture of oligomers and polymers.

Photo-polymerization behaviors were monitored by detecting anodic photocurrent from the TiO_2 electrode. Figure 2 shows chronoamperometric data during PBT formation inside a TiO_2 film with the same polymerization condition as noted in Figure 1. If all collected charges are used for the polymerization, the data implies that the polymerization gradually retards. The right figure of Figure 2 compares linear sweep voltammograms of the TiO_2 film under light irradiation and in dark. In dark, using the BTC/TiO_2 film, the current onset is approximately +1.0 V, corresponding to electrochemical polymerization. In contrast, anodic current onset was recognized at a significantly negative potential of -0.3 V under the light irradiation, supporting that the polymerization is initiated by the BTC excitation followed by the electron injection process from the BTC to the TiO_2.

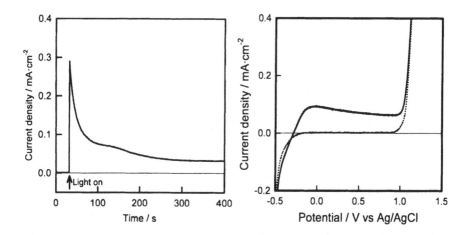

Figure 2. Left: anodic photocurrent profile for the PBT photo-polymerization inside the BTC/TiO_2 nanopores. Right: linear sweep voltammograms of BTC/TiO_2 electrode with (solid line) and without (dotted line) light irradiation. The voltage was applied from -0.5 V to +1.5 V at 50 mV/s.

Influence of the photo-polymerization time on absorption spectral changes was investigated to understand the polymerization behavior. Figure 3a exhibits spectral change as a function of collected electron charges, 8, 25.2, 32.8 and 138.8 mC/cm^2. The absorbance clearly increases with the light irradiation time. Absorption difference from $PBT-BTC/TiO_2$ to BTC/TiO_2 was calculated in order to identify the influence of the polymerization degree on the absorption spectra as shown in Figure 3b. After the photogenerated charges of 8 mC/cm^2 were collected, the difference spectrum indicates a absorption peak at 440 nm, typically observed for oligothiophenes.[28, 40] As the collected charges increase, the peak of the difference spectrum gradually shifts to a longer wavelength to 470 nm, indicative of the polymer formation.[28, 40] The absorption difference was further analyzed at 480 and 550 nm, see Figure 3c. These results clearly show that the relationship is not linear, indicating that some of the photogenerated charges were not consumed for the polymerization reactions. Similar behaviors were observed when a series of thiophene monomers were used as reactants for the photo-polymerization.

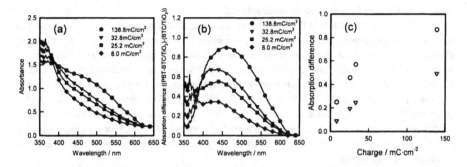

Figure 3. (a) Absorption spectral change of the PBT-BTC/TiO$_2$ as a function of charges collected during the photo-polymerization. (b) Absorption spectral difference, (PBT-BTC/TiO$_2$) – (BTC/TiO$_2$). (c) Absorption difference, (PBT-BTC/TiO$_2$) – (BTC/TiO$_2$), at 480 (circle) and 550 (triangle) nm.

The morphology of the TiO$_2$ films was compared prior to and after the photo-polymerization. Figure 4 shows FE-SEM images of the TiO$_2$ film alone and the PBT-BTC/TiO$_2$ obtained in PC containing 0.1 M LiClO$_4$ for 60 min, corresponding to the collected charges of 63.1 mC/cm^2. Before the polymerization, the TiO$_2$ nanoparticles (diameter: 10~20 nm) are distinctively observed. In contrast, a smooth surface was revealed for the PBT-BTC/TiO$_2$ (Figure 4b) and in the magnified view (Figure 4d), the pore is almost filled by the generated polymer. The cross section image of the PBT-BTC/TiO$_2$ (Figure 4e) exhibits homogeneous polymer formation, suggesting that the photo-polymerization occurs uniformly on nanometer scale.

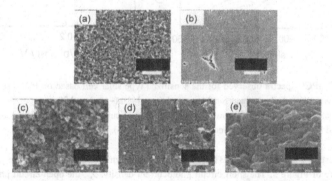

Figure 4. Top FE-SEM images of a TiO$_2$ film (a, c) and PBT-BTC/TiO$_2$ (b, d). A cross section image of PBT-BTC/TiO$_2$ is also shown (e). Magnification: (a) ×100,000, (b) ×40,000 and (c, d, e) ×250,000.

Nanohybrid film based photoelectrochemical solar cells

The prepared nanohybrid film was readily applied as a working electrode in a sensitized-type solar cell. This experimentation was devised to substantiate if this nanohybrid fabrication method can directly be employed for device applications. Figure 5 shows IPCE spectra; the nanohybrid film was prepared in 0.1 M 3,3'-dimethyl-2,2'-bithiophene in BMITFSI under light irradiation at various times. The BTC/TiO$_2$ film clearly exhibits enhanced photocurrent response in a wavelength range of 400~650 nm, with the maximum IPCE of 25% at 400 nm Within 1 min. of the light irradiation, the spectral shape significantly changed. This implies that the IPCE relates to the photo-polymerization degree. Further continuous light irradiation resulted in a gradual decrease in the IPCE at wavelengths shorter than 520 nm. In contrast, at >530 nm, the IPCE initially increases with time and decreases after the excess polymerization time.

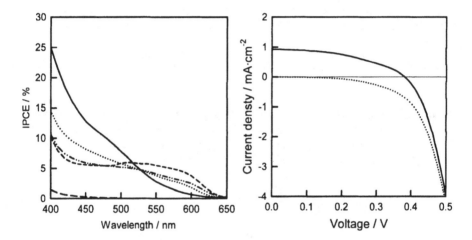

Figure 5. Left: IPCE spectra observed for the sensitized type solar cell based on PDMBT-BTC/TiO$_2$. The PDMBT was photo-polymerized for 0 (——), 1 (······), 10 (– – –) and 30 min. (–··–··–). The IPCE obtained for the TiO$_2$ alone is also shown (— —). Right: J-V curves in dark (······) and under AM1.5 solar simulated light with the light power of 100 mW/cm^2 (——).

J-V characteristics obtained for PDMBT-BTC/TiO$_2$, photopolymerized for 10 min., are presented in the right figure of Figure 5. The short circuit photocurrent density, J_{sc}, the open circuit photovoltage, V_{oc}, the fill factor, FF, and the solar-to-electric conversion efficiency, η, are 0.92 mA/cm^2, 0.38 V, 0.46 and 0.16 %, respectively. For the BTC/TiO$_2$ solar cells, J_{sc}, V_{oc}, FF and η, are 0.70 mA/cm^2, 0.39 V, 0.53 and 0.14 %, respectively. On comparison with the absorbance of the film shown in Figure 1, the electron transfer yield is low compared to the ruthenium dye sensitized solar cells.[36, 41]

Factors influencing photon-to-electron conversion efficiency

The potential energy diagram for the TiO_2 conduction band, the BTC and the I_3^-/I^- redox potentials are summarized in Scheme 1. The TiO_2 conduction band edge and the BTC redox potential were determined experimentally in the previous study.[30] Interestingly, the BTC indicates a wide distribution of redox potentials when adsorbed on the TiO_2 surface, probably owing to local electric fields. Due to this potential distribution, there is much difficulty in determining the LUMO potential. Nevertheless, if the HOMO-LUMO energy difference of the BTC is assumed to 2.3 eV, deduced from the absorption onset (at 540 nm), the lowest LUMO level can provisionally be calculated to ~-0.5 V if we assume the lowest HOMO level to be +1.8 V vs. Ag/AgCl. Comparing the potential energy levels between the TiO_2 conduction band and the BTC excited state potentials, an efficient electron injection from the BTC into the TiO_2 can be expected. The BTC cation state, following the injection process, can readily be re-reduced by the I_3^-/I^- electrolyte.

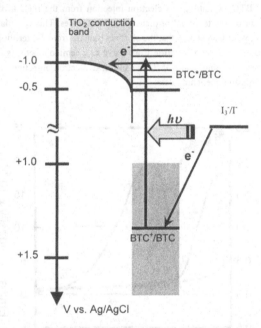

Scheme 1. Potential energy diagram of the TiO_2 electronic states, the adsorbed BTC and the I_3^-/I^- electrolyte. The distribution of the BTC redox potentials is shown according to our recent published results.[30]

There are two key interfacial electron transfer steps influencing photon-to-electron conversion efficiency: from polythiophene to TiO_2 and from an redox electrolyte to oxidized polythiophene. The electron transfer efficiency of the former step was investigated by monitoring emission spectra of the

BTC adsorbed on the TiO_2 and the Al_2O_3. We have employed the BTC/TiO_2 as a model sample to estimate the upper limit of the electron injection efficiency. Since Al_2O_3 is known as an insulator, no electron acceptor state was available for the BTC. Figure 6 exhibits the resultant absorption and emission spectra (355 nm excitation). Comparison of the BTC/Al_2O_3 with the spectrum of the Al_2O_3 film alone indicates that an absorption maximum for BTC is located at approximately 320 nm, and the strong emission with a peak wavelength at 425 nm was observed. In contrast, the emission for the BTC/TiO_2 was completely quenched, suggesting the efficient electron injection from the BTC to the TiO_2. Excitation at 290, 400, 425 and 450 nm also results in complete emission quenching, implying that the TiO_2 excitation at <400 nm does not influence the quenching. The electron injection yield is estimated to be >99 % by calculation from these quenching data, in accordance with the previous reports.[42, 43] We note that yellow coloration of the film (see Figure 6) appears for the TiO_2 only, not for the Al_2O_3.

Following the BTC excitation, the electron injection from the BTC to the TiO_2 conduction band is expected to occur on femtosecond or picosecond time scales. This speculation can be justified by the report published by Janssen et al.[42] that the electron transfer from the terthiophene attached TiO_2 nanoparticle was <4 ps. The electron injection from a dye to a semiconductor is reported to occur on femtosecond and picosecond time scales.[44] These ultrafast electron injection studies are in sound agreement with these emission quenching data.

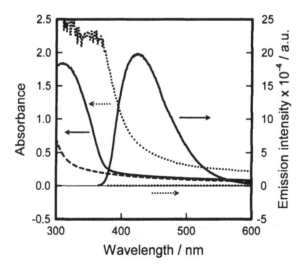

Figure 6. Absorption and emission spectra for BTC adsorbed Al_2O_3 (——) and TiO_2 films (·······). Both films were excited at 355 nm for the emission data. An absorption spectrum of an Al_2O_3 film (– – –) is also shown as comparison.

The BTC regeneration process was observed using a transient absorption spectrometer. Following the excitation, an instrument response limited absorption spectrum was appeared with a peak at 670 nm, as shown in Figure 7. This absorption is provisionally attributed to the BTC cation band since the injected electrons have low absorption coefficients in the range of visible to near infra-red wavelengths. The left figure of Figure 7 compares the cation absorption decays of the BTC/TiO$_2$ films in the presence and the absence of I$^-$. The BTC/TiO$_2$ in the LiClO$_4$ solution indicates charge recombination reactions between the BTC cation and the electron in the TiO$_2$, being attributed to a geminate recombination reaction.[34] The observed recombination with a decay half time of 70 μs is faster than the ruthenium dye/TiO$_2$ (~1 ms).[34, 45] In contrast, slight acceleration of the decay profile was observed for the BTC/TiO$_2$ film in the LiI solution with a half time of 20 μs. This acceleration probably results from the re-reduction of the BTC by I$^-$, competing with the charge recombination process. However, over this wide range of time scales, these similar decay dynamics of the BTC regeneration and the charge recombination clarify the origin of the lower IPCEs observed as in Figure 5. Palomares reported[46] that the distance between the dye and the TiO$_2$ largely influences charge recombination rates for the dye sensitized TiO$_2$ films; shorter the distance, the faster the charge recombination rate. A bithiophene unit is directly linked to the TiO$_2$ through the carboxylic moiety, indicating that the relatively fast charge recombination rate is expected. This study therefore reinforces the importance of the distance control between the chromophore and the TiO$_2$ surface, probably being the key parameter determining the performance of the solar cells based upon the polythiophene/TiO$_2$ nanohybrid films.

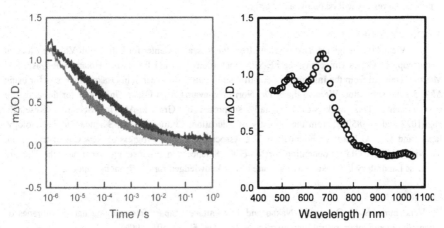

Figure 7. Left: transient absorption decays of the BTC cation states at 650 nm, obtained for the BTC/TiO$_2$ film in the presence of 0.1 M LiClO$_4$ (black line) and 0.1 M LiI (red line) in the solvent with the light excitation at 425 nm. The absorption decay half times are 70 and 20 μs for 0.1 M LiClO$_4$ and 0.1 M LiI, respectively. Right: a transient spectrum of the charge separated state (BTC$^+$ and e$^-$(TiO$_2$)) at 1 μs after the excitation.

CONCLUSION

We have conducted photoinduced polymerization of thiophene inside the TiO_2 nanoporous film. Initially, thiophene monomers possessing a carboxyl group, 2,2'-bithiophene-5-carboxylic acid (BTC), were adsorbed on the TiO_2 surface, and a selective excitation of the BTC led to the thiophene photo-polymerization inside the TiO_2 pore. Thus, this method has thoroughly provided nanohybridization and electronic connection within TiO_2.

The photocurrent generated in the solar cell is largely dependent on the photo-polymerization degree. After the light irradiation for 1 min., the IPCE spectrum is remarkably different from the BTC/TiO_2 film. This drastic change is expected since the photo-polymerization rapidly occurs within 1 min. and the polymer grows vertically from the TiO_2 surface. The IPCE decreases with the light irradiation time although the light harvesting efficiency, light absorption by the film, improves after the photo-polymerization.

The interfacial kinetic studies have revealed the high electron injection efficiency (>99 %) from the surface attached thiophene molecule to the TiO_2. In contrast, the charge recombination between the electron in the TiO_2 and the BTC oxidized state has exhibited similar multi-exponential kinetics to those of the charge regeneration reaction for the BTC cation. This kinetic comparison clarifies that the photon-to-electric conversion efficiency is limited by a slow charge regeneration process or a fast charge recombination process. This relative kinetic difference will be a key factor to improve the efficiency. We are currently developing a novel conducting polymer/metal oxide nanohybrid system by controlling molecular structure of the conducting polymer. The detailed experimental results will be published shortly.

ACKNOWLEDGEMENTS

We acknowledge Dr. Takao Sakata from the Research Center for Ultra-High Voltage Electron Microscopy in Osaka University for FE-SEM measurements, and Professors Hikaru Kobayashi and Masao Takahashi from the Institute of Scientific and Industrial Research in Osaka University for using AM1.5 solar simulator. We also thank Dr. Norio Nagayama from Osaka University for the emission measurements. This work was financially supported by Grant-in-Aid for Scientific Research, 18201022 and 18685002, from the Ministry of Education, Culture, Sports, Science and Technology, Japan, and TEPCO Research Foundation. The Association for the Progress of New Chemistry, Japan, the Nippon Sheet Glass Foundation for Materials Science and Engineering, Japan and the Venture Business Laboratory in Osaka University are also acknowledged for the financial support.

REFERENCES

[1]R. Vendamme, S.-Y. Onoue, A. Nakao and T. Kunitake, Robust free-standing nanomembranes of organic/inorganic interpenetrating networks, *Nat. Mater.*, **5**, 494-501 (2006).
[2]K. Valle, P. Belleville, F. Pereira and C. Sanchez, Hierarchically structured transparent hybrid membranes by in situ growth of meso-structured organosilica in host polymer, *Nat. Mater.*, **5**, 107-11 (2006).
[3]Y. Lin, A. Boeker, J. He, K. Sill, H. Xiang, C. Abetz, X. Li, J. Wang, T. Emrick, S. Long, Q. Wang, A. Balazs and T. P. Russell, Self-directed self-assembly of nanoparticle/copolymer mixtures, *Nature*

(London), **434**, 55-59 (2005).

[4]L. Li, E. Beniash, E. R. Zubarev, W. Xiang, B. M. Rabatic, G. Zhang and S. I. Stupp, Assembling a lasing hybrid material with supramolecular polymers and nanocrystals, *Nat. Mater.*, **2**, 689-94 (2003).

[5]S. Moeller, C. Perlov, W. Jackson, C. Taussig and S. R. Forrest, A polymer/semiconductor write-once read-many-times memory, *Nature (London)*, **426**, 166-69 (2003).

[6]K. Tajima, L.-S. Li and S. I. Stupp, Nanostructured Oligo(p-phenylene Vinylene)/Silicate Hybrid Films: One-Step Fabrication and Energy Transfer Studies, *J. Am. Chem. Soc.*, **128**, 5488-95 (2006).

[7]Y. Cao, I. D. Parker, G. Yu, C. Zhang and A. J. Heeger, Improved quantum efficiency for electroluminescence in semiconducting polymers, *Nature (London)*, **397**, 414-17 (1999).

[8]N. Tessler, G. J. Denton and R. H. Friend, Lasing from conjugated-polymer microcavities, *Nature (London)*, **382**, 695-97 (1996).

[9]N. Stutzmann, R. H. Friend and H. Sirringhaus, Self-Aligned, Vertical-Channel, Polymer Field-Effect Transistors, *Science (Washington, D.C.)*, **299**, 1881-85 (2003).

[10]L. Schmidt-Mende, A. Fechtenkotter, K. Mullen, E. Moons, R. H. Friend and J. D. MacKenzie, Self-organized discotic liquid crystals for high-efficiency organic photovoltaics, *Science (Washington, D.C.)*, **293**, 1119-22 (2001).

[11]H. Kawazoe, M. Yasukawa, H. Kyodo, M. Kurita, H. Yanagi and H. Hosono, P-type electrical conduction in transparent thin films of $CuAlO_2$, *Nature (London)*, **389**, 939-42 (1997).

[12]S.-H. Lee, R. Deshpande, P. A. Parilla, K. M. Jones, B. To, H. Mahan and A. C. Dillon, Crystalline WO_3 nanoparticles for highly improved electrochromic applications, *Adv. Mater. (Weinheim, Ger.)*, **18**, 763-66 (2006).

[13]A. Fujishima and K. Honda, Electrochemical photolysis of water at a semiconductor electrode, *Nature (London)*, **238**, 37-38 (1972).

[14]B. O' Regan and M. Grätzel, A low-cost, high-efficiency solar cell based on dye sensitized colloidal TiO_2 films, *Nature (London)*, **353**, 737-39 (1991).

[15]Y. Liu, M. A. Summers, C. Edder, J. M. J. Frechet and M. D. McGehee, Using resonance energy transfer to improve exciton harvesting in organic-inorganic hybrid photovoltaic cells, *Adv. Mater. (Weinheim, Ger.)*, **17**, 2960-64 (2005).

[16]W. J. E. Beek, M. M. Wienk and R. A. J. Janssen, Hybrid solar cells from regioregular polythiophene and ZnO nanoparticles, *Adv. Funct. Mater.*, **16**, 1112-16 (2006).

[17]P. Ravirajan, S. A. Haque, J. R. Durrant, D. Poplavskyy, D. D. C. Bradley and J. Nelson, Hybrid nanocrystalline TiO_2 solar cells with a fluorene-thiophene copolymer as a sensitizer and hole conductor, *J. Appl. Phys.*, **95**, 1473-80 (2004).

[18]P. Ravirajan, A. M. Peiro, M. K. Nazeeruddin, M. Grätzel, D. D. C. Bradley, J. R. Durrant and J. Nelson, Hybrid Polymer/Zinc Oxide Photovoltaic Devices with Vertically Oriented ZnO Nanorods and an Amphiphilic Molecular Interface Layer, *J. Phys. Chem. B*, **110**, 7635-39 (2006).

[19]C. L. Huisman, A. Goossens and J. Schoonman, Aerosol Synthesis of Anatase Titanium Dioxide Nanoparticles for Hybrid Solar Cells, *Chem. Mater.*, **15**, 4617-24 (2003).

[20]A. C. Arango, L. R. Johnson, V. N. Bliznyuk, Z. Schlesinger, S. A. Carter and H.-H. Horhold, Efficient titanium oxide/conjugated polymer photovoltaics for solar energy conversion, *Adv. Mater. (Weinheim, Ger.)*, **12**, 1689-92 (2000).

[21]L. H. Slooff, M. M. Wienk and J. M. Kroon, Hybrid TiO_2:polymer photovoltaic cells made from a titanium oxide precursor, *Thin Solid Films*, **451-452**, 634-38 (2004).

[22]C. Y. Kwong, A. B. Djurisic, P. C. Chui, K. W. Cheng and W. K. Chan, Influence of solvent on film morphology and device performance of poly(3-hexylthiophene):TiO_2 nanocomposite solar cells, *Chem. Phys. Lett.*, **384**, 372-75 (2004).

[23]C. D. Grant, A. M. Schwartzberg, G. P. Smestad, J. Kowalik, L. M. Tolbert and J. Z. Zhang, Characterization of nanocrystalline and thin film TiO_2 solar cells with poly(3-undecyl-2,2'-bithiophene) as a sensitizer and hole conductor, *J. Electroanal. Chem.*, **522**, 40-48 (2002).

[24]K. M. Coakley and M. D. McGehee, Photovoltaic cells made from conjugated polymers infiltrated into mesoporous titania, *Appl. Phys. Lett.*, **83**, 3380-82 (2003).

[25]A. F. Nogueira, N. Alonso-Vante and M.-A. De Paoli, Solid-state photoelectrochemical device using poly(o-methoxy aniline) as sensitizer and an ionic conductive elastomer as electrolyte, *Synth. Met.*, **105**, 23-27 (1999).

[26]J. Liu, E. N. Kadnikova, Y. Liu, M. D. McGehee and J. M. J. Frechet, Polythiophene Containing Thermally Removable Solubilizing Groups Enhances the Interface and the Performance of Polymer-Titania Hybrid Solar Cells, *J. Am. Chem. Soc.*, **126**, 9486-87 (2004).

[27]R. Senadeera, N. Fukuri, Y. Saito, T. Kitamura, Y. Wada and S. Yanagida, Volatile solvent-free solid-state polymer-sensitized TiO_2 solar cells with poly(3,4-ethylenedioxythiophene) as a hole-transporting medium, *Chem. Commun.*, 2259-61 (2005).

[28]C. L. Huisman, A. Huijser, H. Donker, J. Schoonman and A. Goossens, UV polymerization of oligothiophenes and their application in nano structured heterojunction solar cells, *Macromolecules*, **37**, 5557-64 (2004).

[29]Y.-G. Kim, J. Walker, L. A. Samuelson and J. Kumar, Efficient Light Harvesting Polymers for Nanocrystalline TiO_2 Photovoltaic Cells, *Nano Lett.*, **3**, 523-25 (2003).

[30]Y. Otsuka, Y. Okamoto, H. Y. Akiyama, K. Umekita, Y. Tachibana and S. Kuwabata, Photoinduced Formation of Polythiophene/TiO_2 Nanohybrid Heterojunction Films for Solar Cell Applications, *J. Phys. Chem. C*, **112**, 4767-75 (2008).

[31]Y. Tachibana, R. Muramoto, H. Matsumoto and S. Kuwabata, Photoelectrochemistry of p-type Cu_2O semiconductor electrode in ionic liquid, *Res. Chem. Intermed.*, **32**, 575-83 (2006).

[32]H. Matsumoto, H. Kageyama and Y. Miyazaki, Effect of ionic additives on the limiting cathodic potential of EMI-based room temperature ionic liquids, *Electrochemistry*, **71**, 1058-60 (2003).

[33]G. Redmond and D. Fitzmaurice, Spectroscopic Determination of Flatband Potentials for Polycrystalline TiO_2 Electrodes in Nonaqueous Solvents, *J. Phys. Chem.*, **97**, 1426-30 (1993).

[34]S. A. Haque, Y. Tachibana, R. L. Willis, J. E. Moser, M. Grätzel, D. R. Klug and J. R. Durrant, Parameters Influencing Charge Recombination Kinetics in Dye-Sensitized Nanocrystalline Titanium Dioxide Films, *J. Phys. Chem. B*, **104**, 538-47 (2000).

[35]I. Montanari, J. Nelson and J. R. Durrant, Iodide Electron Transfer Kinetics in Dye-Sensitized Nanocrystalline TiO_2 Films, *J. Phys. Chem. B*, **106**, 12203-10 (2002).

[36]Y. Tachibana, K. Hara, K. Sayama and H. Arakawa, Quantitative Analysis of Light-Harvesting Efficiency and Electron-Transfer Yield in Ruthenium-Dye-Sensitized Nanocrystalline TiO_2 Solar Cells,

Chem. Mater., **14**, 2527-35 (2002).

[37]T. Rajh, L. X. Chen, K. Lukas, T. Liu, M. C. Thurnauer and D. M. Tiede, Surface Restructuring of Nanoparticles: An Efficient Route for Ligand-Metal Oxide Crosstalk, *J. Phys. Chem. B*, **106**, 10543-52 (2002).

[38]G. Ramakrishna, S. Verma, D. A. Jose, D. K. Kumar, A. Das, D. K. Palit and H. N. Ghosh, Interfacial Electron Transfer between the Photoexcited Porphyrin Molecule and TiO$_2$ Nanoparticles: Effect of Catecholate Binding, *J. Phys. Chem. B*, **110**, 9012-21 (2006).

[39]G. Tourillon and F. Garnier, New electrochemically generated organic conducting polymers, *J. Electroanal. Chem.*, **135**, 173-8 (1982).

[40]D. Fichou and Editor, Handbook of Oligo- and Polythiophenes, 534 pp. (1999).

[41]M. K. Nazeeruddin, P. Pé chy, T. Renouard, S. M. Zakeeruddin, R. Humphry-Baker, P. Comte, P. Liska, L. Cevey, E. Costa, V. Shklover, L. Spiccia, G. B. Deacon, C. A. Bignozzi and M. Grätzel, Engineering of efficient panchromatic sensitizers for nanocrystalline TiO$_2$-based solar cells, *J. Am. Chem. Soc.*, **123**, 1613-24 (2001).

[42]W. J. E. Beek and R. A. J. Janssen, Spacer length dependence of photoinduced electron transfer in heterosupramolecular assemblies of TiO$_2$ nanoparticles and terthiophene, *J. Mater. Chem.*, **14**, 2795-800 (2004).

[43]X. Ai, N. Anderson, J. Guo, J. Kowalik, L. M. Tolbert and T. Lian, Ultrafast Photoinduced Charge Separation Dynamics in Polythiophene/SnO$_2$ Nanocomposites, *J. Phys. Chem. B*, **110**, 25496-503 (2006).

[44]Y. Tachibana, S. A. Haque, I. P. Mercer, J. R. Durrant and D. R. Klug, Electron Injection and Recombination in Dye Sensitized Nanocrystalline Titanium Dioxide Films: A Comparison of Ruthenium Bipyridyl and Porphyrin Sensitizer Dyes, *J. Phys. Chem. B*, **104**, 1198-205 (2000).

[45]Y. Tachibana, J. E. Moser, M. Grätzel, D. R. Klug and J. R. Durrant, Subpicosecond interfacial charge separation in dye-sensitized nanocrystalline titanium dioxide films, *J. Phys. Chem.*, **100**, 20056-62 (1996).

[46]E. Palomares, J. N. Clifford, S. A. Haque, T. Lutz and J. R. Durrant, Control of Charge Recombination Dynamics in Dye Sensitized Solar Cells by the Use of Conformally Deposited Metal Oxide Blocking Layers, *J. Am. Chem. Soc.*, **125**, 475-82 (2003).

LIQUID PHASE MORPHOLOGY CONTROL OF METAL OXIDES—PHASE TRANSFORMATION OF STAND-ALONE ZnO FILMS IN AQUEOUS SOLUTIONS

Yoshitake Masuda
National Institute of Advanced Industrial Science and Technology (AIST), 2266-98 Anagahora, Shimoshidami, Moriyama-ku, Nagoya 463-8560, Japan
* Corresponding Author: Y. Masuda, masuda-y@aist.go.jp

ABSTRACT

Stand-alone ZnO films were prepared from aqueous solutions. They were formed at air-liquid interface, i.e., top of the solutions which were used as templates. The films were assemblies of ZnO nanosheets. Cross section profiles of the films showed gradient structures consisted of the nanosheets and nanopores. The films had smooth air side surfaces and rough liquid side surfaces. They showed high c-axis orientation in XRD analyses. The films can also be pasted onto desired substrates such as polymer films or mesh substrates. The films had many unique properties and can be applied to various next-generation devices. Furthermore, stand-alone ZnO films were showed to be phase transform to $Zn_5(CO_3)_2(OH)_6$ in the solutions after 1 month. The films were assemblies of nanosheets of $Zn_5(CO_3)_2(OH)_6$. ZnO films gradually transformed to $Zn_5(CO_3)_2(OH)_6$ films in the solutions with maintenance of their structures.

INTRODUCTION

ZnO is an attractive oxide for dye-sensitised solar cells[1-3], gas sensors[4,5], varistors[6], piezoelectric devices[7], electroacoustic transducers[8], vacuum fluorescent displays[9], field emission displays[10,11], electroluminescent displays[12], UV light-emitting diodes, laser diodes[13-15], and displays[16,17]. Sensitivity directly depends on the specific surface area of sensing materials. ZnO particles, particulate films or mesoporous material having high specific surface area are thus strongly required. Additionally, high c-axis oriented ZnO films have attractive properties such as photocurrent phenomena[18] and are expected to be used for future optical and electrical devices.

However, ZnO films have usually been prepared on substrates[19-37], and in particular crystalline ZnO films having high c-axis orientation require expensive substrates such as single crystals or highly-functional substrates. A simple and low-cost process for self-supporting crystalline ZnO films is expected to be used for a wide range of applications such as windows of optical devices or low-value-added products. Self-supporting crystalline ZnO films can also be applied by being pasting on a desired substrate such as low heat-resistant polymer films, glasses, metals or papers.

As a precursor of ZnO[38], zinc carbonate hydroxide ($Zn_5(CO_3)_2(OH)_6$) has attracted much attention. The morphology of ZnO can be controlled by the crystal growth of $Zn_5(CO_3)_2(OH)_6$[38], and controlling the morphology of ZnO and $Zn_5(CO_3)_2(OH)_6$ films is very important for the development of ZnO-based devices[39].

In this study, stand-alone ZnO films were formed at air-liquid interface of the solutions[40]. The films were assemblies of ZnO nanosheets having unique properties. Furthermore, the films gradually transformed to $Zn_5(CO_3)_2(OH)_6$ films in the solutions with maintenance of their structures[41].

EXPERIMENT

Zinc nitrate hexahydrate ($Zn(NO_3)_2 \cdot 6H_2O$, > 99.0%, MW 297.49, Kanto Chemical Co., Inc.) and ethylenediamine ($H_2NCH_2CH_2NH_2$, > 99.0%, MW 60.10, Kanto Chemical Co., Inc.) were used as received. Zinc nitrate hexahydrate (15 mM) was dissolved in distilled water at 60°C and ethylenediamine (15 mM) was added to the solutions to induce the formation of ZnO. The solutions were kept at 60°C using a water bath for 6 h with no stirring. The solutions were then left to cool for 42 h or 1 month in the bath. Polyethylene terephthalate (PET) films, glasses (S-1225,

Matsunami Glass Ind., Ltd.) and Si wafers (p-type Si [100], NK Platz Co., Ltd.) were used as substrates.

Morphology of ZnO films were observed by a field emission scanning electron microscope (FE-SEM; JSM-6335FM, JEOL Ltd.) and a transmission electron microscope (TEM; H-9000UHR, 300 kV, Hitachi). Crystal phases were evaluated by an X-ray diffractometer (XRD; RINT-2100V, Rigaku) with CuKα radiation (40 kV, 40 mA). Si wafers were used as substrates for XRD evaluation. The crystal structure model and diffraction pattern of ZnO were calculated from ICSD (Inorganic Crystal Structure Database) data No. 26170 (FIZ Karlsruhe, Germany and NIST, USA) using FindIt and ATOMS (Hulinks Inc.).

RESULTS AND DISCUSSION

Stand-alone ZnO film[40]

The solution became clouded shortly after the addition of ethylenediamine by the homogeneous nucleation and growth of ZnO particles. Ethylenediamine plays an essential role in the formation of crystalline ZnO. Zinc-ethylenediamine complex forms in the solution as follows[42]:

$$Zn^{2+} + 3NH_2 \cdot (CH_2)_2 \cdot NH_2 \rightleftharpoons [Zn(NH_2 \cdot (CH_2)_2 \cdot NH_2)_3]^{2+} \qquad (1)$$

The chemical equilibrium in eq. 1 moves to the left and the zinc-ethylenediamine complex decomposes, causing the concentration of Zn^{2+} to increase at elevated temperature.

OH^- concentration increases by the hydrolysis of ethylenediamine as follows:

$$NH_2 \cdot (CH_2)_2 \cdot NH_2 + 2H_2O \rightleftharpoons NH_3 \cdot (CH_2)_2 \cdot NH_3^{2+} + 2OH^- \qquad (2)$$

ZnO and $Zn(OH)_2$ are thus formed in the aqueous solution as follows:

$$Zn^{2+} + 2OH^- \longrightarrow Zn(OH)_2 \longrightarrow ZnO + H_2O \qquad (3)$$

ZnO particles were gradually deposited and covered the bottom of the vessels, and the solutions became light white after 1 h and clear after 6 h. The supersaturation degree of the solutions was high at the initial stage of the reaction for the first 1 h and decreased as the color of the solution schanged.

White films were formed at the air-liquid interface and they grew into large films. The films had sufficiently high strength to be self-supporting. Additionally, films were pasted onto desired substrates such PET films, Si wafers, glass plates or papers, and the pasted ZnO films were then dried to bond it to the substrates.

The film showed a very strong 0002 x-ray diffraction peak of hexagonal ZnO at $2\theta = 34.04°$ and weak 0004 diffraction peak at $2\theta = 72.16°$ with no other diffractions of ZnO (Fig. 1a). (0002) planes and (0004) planes were perpendicular to the c-axis, and the diffraction peak only from (0002) and (0004) planes indicates high c-axis orientation of ZnO film. The inset figure shows that the crystal structure of hexagonal ZnO stands on a substrate to make the c-axis perpendicular to the substrate. Crystallite size parallel to (0002) planes was estimated from the half-maximum full-width of the 0002 peak to 43 nm. This is similar to the threshold limit value of our XRD equipment and thus the crystallite size parallel to (0002) planes is estimated to be greater than or equal to 43 nm. Diffraction peaks from a silicon substrate were observed at $2\theta = 68.9°$ and $2\theta = 32.43°$. Weak diffractions at $2\theta = 12.5°, 24.0°, 27.6°, 30.5°, 32.4°$ and $57.6°$ were assigned to co-precipitated zinc carbonate hydroxide ($Zn_5(CO_3)_2(OH)_6$, JCPDS No. 19-1458).

The film grew to a thickness of about 5 μm after 48 h, i.e., 60°C for 6 h, and was left to cool for 42 h.

The air side of the stand-alone film had a smooth surface over a wide area due to the flat air-liquid interface (Fig. 2), whereas the liquid side of the film had a rough surface. The films consisted of ZnO nano-sheets were clearly observed from the liquid side and the fracture edge-on profile of the film. The nano-sheets had a thickness of 5-10 nm and were 1-5 μm in size. They mainly grew forward to the bottom of the solution, i.e., perpendicular to the air-liquid interface, such that the sheets stood perpendicular to the air-liquid interface. Thus, the liquid side of the film had many ultra-fine spaces surrounded by nano-sheet and had a high specific surface area. The air side of the film, on the other hand, had a flat surface that followed the flat shape of the air-liquid interface. The air-liquid interface was thus effectively utilized to form the flat surface of the film. This flatness would contribute to the strong adhesion strength to substrates for pasting of the film. The air-side surface prepared for 48 h had holes of 100-500 nm in diameter, and were hexagonal, rounded hexagonal or round in shape. The air-side surface prepared for 6 h, in contrast, had no holes on the surface. The air-side surface was well crystallized to form a dense surface and ZnO crystals would partially grow to a hexagonal shape because of the hexagonal crystal structure. Well-crystallized ZnO hexagons were then etched to form holes on the surface by decrease in pH. The growth face of the film would be liquid side. ZnO nano-sheets would grow to form a large ZnO film by Zn ion supply from the aqueous solution. Additionally, holes were not observed from air-side surface near edges of the films (Fig. 2a, b). The films become larger over time during immersion period. Edges of the films grew after formation of center of the films. It supported the formation mechanism of holes mentioned above.

The film pasted on a silicon wafer was annealed at 500°C for 1 h in air to evaluate the details of the films. ZnO film maintained its structure during the annealing (Fig. 3). The air side of the film showed a smooth surface and the liquid side showed a relief structure having a high specific surface area. The air side showed the film consisted of dense packing of small ZnO nanosheets and the size of sheets increased toward the liquid-side surface. ZnO sheets would grow from the air side to the liquid side, i.e., the sheets would nucleate at the liquid-air interface and grow down toward the bottom of the solution by the supply of Zn ions from the solution. Annealed film showed X-ray diffractions of ZnO and Si substrate with no additional phases. As-deposited ZnO nano-sheets were shown to be crystalline ZnO because the sheets maintained their fine structure during the annealing without any phase transition. High c-axis orientation was also maintained during the annealing, showing a very strong 0002 diffraction peak.

Stand-alone ZnO film was further evaluated by TEM and electron diffraction (Fig. 4). The film was crushed to sheets and dispersed in an acetone. The sheets at the air-liquid interface were skimmed by a cupper grid with a carbon supporting film. The sheets were shown to have uniform thickness. They were dense polycrystalline films constructed of ZnO nanoparticles. Lattice image was clearly observed to show high crystallinity of the particles. The film was shown to be single phase of ZnO by electron diffraction pattern. These observations were consistent with XRD and SEM evaluations.

The films had several features suitable for sensing applications and dye-sensitized solar cells. They had nano/micro surface relief structures and high surface area in liquid side surfaces. Sensing properties and dye adsorption properties are directly affected by these features. The films were consisted of nano-sheets stood perpendicular to the films. The nano-sheets had uniform thickness and were dense ZnO crystals. The nano-sheets had no grain boundaries which decrease electrical conductivity as seen in common sintered particulate films. The films had high conductivity perpendicular to the films. The feature is important factor for sensing applications and dye-sensitized solar cells. Moreover, the films had high optical transparency. It increases photoelectric conversion efficiency of solar cells and dye-sensitized sensors[43]. The films are candidate material for future sensing applications and dye-sensitized solar cells.

Stand-alone $Zn_5(CO_3)_2(OH)_6$ film

The solution was further kept at 25 °C for 1 month. The film formed at the air-liquid interface had sufficiently high strength to be self-supporting. It was pasted onto a glass substrate for XRD and FE-SEM evaluation. X-ray diffraction peaks were observed at 2θ = 13.0 °, 24.1 °, 27.9 °, 30.9 °, 32.7 °, 35.6 °, 38.6 °, 47.0 °, 47.2 ° and 54.3 ° (Fig. 1b). They were assigned to 200, -310, 020, 220, 021, 510, -420, 222, 330 and 800 diffraction peaks of zinc carbonate hydroxide $(Zn_5(CO_3)_2(OH)_6$[44]) (JCPSD No. 19-1458). Crystallite size was estimated from the peak at 2θ = 13.0 ° to be 18 nm. A broad diffraction peak from the glass substrate was observed at about 2θ = 25 °.

ZnO was crystallized at 60 °C for 6 h. ZnO would be gradually etched and dissolved by nitric acid[45-47] as the solution temperature decreased. $Zn_5(CO_3)_2(OH)_6$ was then crystallized using Zn ions which were supplied by the dissolution of crystalline ZnO.

Morphology of the film was further evaluated by FE-SEM. The air side of the film was smooth all over the surface, reflecting the flatness of the air-liquid interface (Fig. 5). The flatness would contribute to the strong strength with which the pasted film adhered to the substrates. The film was an assembly of thin sheets which were connected with each other. The nano-sheets had a thickness of 5 nm – 20 nm and were 100 nm – 500 nm in size. The surface thus had many pores surrounded by thin sheets.

On the other hand, the liquid side had a rough surface. Additionally, flower-like particles constructed of thin sheets were adhered on the liquid surface. The sheets had a thickness of 5 nm – 30 nm and were 500 nm – 3 µm in size, and were larger than those observed on the air side. This would be because the growth face was the liquid side of the film and the sheets grew downward by the supply of ions from the solution. The liquid surface had a large roughness due to the assembly of large thin sheets.

The film had a smooth surface on the air side and a rough surface on the liquid side. This characteristic is similar to that of ZnO film fabricated for 48 h. Flatness of the air-liquid interface was effectively used to create a smooth surface on the air side of the film. Additionally, both of the films were constructed of thin sheets and the sheets grew downward. The sheets on the liquid side were thus larger than those on the air side or inside of the films. The rough surface of the liquid side depended on the crystal growth in the solution. An air-liquid interface has two different chemical fields: an air side and a liquid side. Both of the conditions were optimized for the development of self-supported films. A smooth surface constructed of a dense assembly of sheets and a rough surface constructed of large sheets were attained in the same film by using the air-liquid interface.

CONCLUSION

Phase transformation of stand-alone ZnO films were investigated in this work. Stand alone ZnO films were firstly prepared at air-liquid interface the aqueous solutions without substrates. The films were consisted of ZnO nanosheets. The films showed high c-axis orientation in XRD analyses. They had smooth air-side surfaces and rough liquid-side surfaces. Gradient structures were observed with cross-sectional SEM observations. The films were further maintained at air-liquid interfaces for 1 month. They gradually transformed to $Zn_5(CO_3)_2(OH)_6$ films in the solutions with maintenance of their structures. Stand alone $Zn_5(CO_3)_2(OH)_6$ films also had smooth air-side surfaces, rough liquid-side surfaces and gradient structures. Phase transformation of metal oxides in aqueous solutions will open up a new field of metal oxide chemistry.

REFERENCES

[1] M. Law, L. E. Greene, J. C. Johnson, R. Saykally, and P. D. Yang, "Nanowire dye-sensitized solar cells," Nature Mater., 4(6), 455-59 (2005).

[2] J. B. Baxter and E. S. Aydil, "Nanowire-based dye-sensitized solar cells," Appl. Phys. Lett., 86(5), 53114 (2005).

[3] S. Karuppuchamy, K. Nonomura, T. Yoshida, T. Sugiura, and H. Minoura, "Cathodic electrodeposition of oxide semiconductor thin films and their application to dye-sensitized solar cells," Solid State Ionics, 151(1-4), 19-27 (2002).

[4] N. Golego, S. A. Studenikin, and M. J. Cocivera, "Sensor photoresponse of thin-film oxides of zinc and titanium to oxygen gas," J. Electrochem. Soc., 147(4), 1592-94 (2000).

[5] G. Sberveglieri, "Recent Developments in Semiconducting Thin-Film Gas Sensors," Sens. Actuators B: Chem., 23(2-3), 103-09 (1995).

[6] Y. Lin, Z. Zhang, Z. Tang, F. Yuan, and J. Li, "Characterisation of ZnO-based varistors prepared from nanometre precursor powders," Adv. Mater. Opt. Electron., 9(5), 205-09 (1999).

[7] G. Agarwal and R. F. Speyer, "Current change method of reducing gas sensing using ZnO varistors," J. Electrochem. Soc., 145(8), 2920-25 (1998).

[8] F. Quaranta, A. Valentini, F. R. Rizzi, and G. Casamassima, "Dual-ion beam sputter-deposition of ZnO films," J. Appl. Phys., 74(1), 244-48 (1993).

[9] S. Ruan, Proc SPIE, 262, 2892 (1996).

[10] Y. Nakanishi, A. Miyake, H. Kominami, T. Aoki, Y. Hatanaka, and G. Shimaoka, "Preparation of ZnO thin films for high-resolution field emission display by electron beam evaporation," Appl. Surf. Sci., 142(1-4), 233-36 (1999).

[11] S. Fujihara, Y. Ogawa, and A. Kasai, "Tunable visible photoluminescence from ZnO thin films through Mg-doping and annealing," Chem. Mater., 16(15), 2965-68 (2004).

[12] L. Yi, Y. Hou, H. Zhao, D. He, Z. Xu, Y. Wang, and X. Xu, "The photo- and electro-luminescence properties of ZnO : Zn thin film," Displays, 21(4), 147-49 (2000).

[13] R. F. Service, "Will UV lasers beat the blues?," Science, 276, 895 (1997).

[14] A. Hellemans, "Physics - Laser light from a handful of dust," Science, 284(5411), 24 (1999).

[15] M. H. Huang, S. Mao, H. Feick, H. Yan, Y. Wu, H. Kind, E. Weber, R. Russo, and P. Yang, "Room-Temperature Ultraviolet Nanowire Nanolasers," Science, 292, 1897-99 (2001).

[16] T. Pauporte and D. Lincot, "Electrodeposition of semiconductors for optoelectronic devices: results on zinc oxide," Electrochem. Acta., 45(20), 3345-53 (2000).

[17] R. Konenkamp, K. Boedecker, M. C. Lux-Steiner, M. Poschenrieder, F. Zenia, C. Levey-Clement, and S. Wagner, "Thin film semiconductor deposition on free-standing ZnO columns," Appl. Phys Lett., 77(16), 2575-77 (2000).

[18] D. H. Zhang and D. E. Brodie, "Crystallite Orientation and the Related Photoresponse of Hexagonal Zno Films Deposited by Rf-Sputtering," Thin Solid Films, 251(2), 151-56 (1994).

[19] S. Yamabi and H. Imai, "Growth conditions for wurtzite zinc oxide films in aqueous solutions," J. Mater. Chem., 12(12), 3773-78 (2002).

[20] K. Kakiuchi, E. Hosono, T. Kimura, H. Imai, and S. Fujihara, "Fabrication of mesoporous ZnO nanosheets from precursor templates grown in aqueous solutions," J. Sol-Gel Sci. Technol., 39(1), 63-72 (2006).

[21] B. Schwenzer, J. R. Gomm, and D. E. Morse, "Substrate-induced growth of nanostructured zinc oxide films at room temperature using concepts of biomimetic catalysis," Langmuir, 22(24), 9829-31 (2006).

[22] N. Saito, H. Haneda, T. Sekiguchi, N. Ohashi, I. Sakaguchi, and K. Koumoto, "Low-Temperature Fabrication of Light-Emitting Zinc Oxide Micropatterns Using

Self-Assembled Monolayers," Adv. Mater., 14(6), 418-21 (2002).

[23] H. D. Yu, Z. P. Zhang, M. Y. Han, X. T. Hao, and F. R. Zhu, "A general low-temperature route for large-scale fabrication of highly oriented ZnO nanorod/nanotube arrays," J. Am. Chem. Soc., 127(8), 2378-79 (2005).

[24] K. S. Choi, H. C. Lichtenegger, G. D. Stucky, and E. W. McFarland, "Electrochemical synthesis of nanostructured ZnO films utilizing self-assembly of surfactant molecules at solid-liquid interfaces," J. Am. Chem. Soc., 124(42), 12402-03 (2002).

[25] M. Yin, Y. Gu, I. L. Kuskovsky, T. Andelman, Y. Zhu, G. F. Neumark, and S. O'Brien, "Zinc oxide quantum rods," J. Am. Chem. Soc., 126(20), 6206-07 (2004).

[26] X. J. Feng, L. Feng, M. H. Jin, J. Zhai, L. Jiang, and D. B. Zhu, "Reversible super-hydrophobicity to super-hydrophilicity transition of aligned ZnO nanorod films," J. Am. Chem. Soc., 126(1), 62-63 (2004).

[27] T. Yoshida, M. Tochimoto, D. Schlettwein, D. Wohrle, T. Sugiura, and H. Minoura, "Self-assembly of zinc oxide thin films modified with tetrasulfonated metallophthalocyanines by one-step electrodeposition," Chem. Mater., 11(10), 2657-67 (1999).

[28] J. Y. Lee, D. H. Yin, and S. Horiuchi, "Site and morphology controlled ZnO deposition on pd catalyst prepared from Pd/PMMA thin film using UV lithography," Chem. Mater., 17(22), 5498-503 (2005).

[29] K. Kopalko, M. Godlewski, J. Z. Domagala, E. Lusakowska, R. Minikayev, W. Paszkowicz, and A. Szczerbakow, "Monocrystalline ZnO films on GaN/Al2O3 by atomic layer epitaxy in gas flow," Chem. Mater., 16(8), 1447-50 (2004).

[30] R. Turgeman, O. Gershevitz, M. Deutsch, B. M. Ocko, A. Gedanken, and C. N. Sukenik, "Crystallization of highly oriented ZnO microrods on carboxylic acid-terminated SAMs," Chem. Mater., 17(20), 5048-56 (2005).

[31] R. Turgeman, O. Gershevitz, O. Palchik, M. Deutsch, B. M. Ocko, A. Gedanken, and C. N. Sukenik, "Oriented growth of ZnO crystals on self-assembled monolayers of functionalized alkyl silanes," Cryst. Growth Des., 4(1), 169-75 (2004).

[32] E. Mirica, G. Kowach, P. Evans, and H. Du, "Morphological evolution of ZnO thin films deposited by reactive sputtering," Cryst. Growth Des., 4(1), 147-56 (2004).

[33] Y. R. Lin, Y. K. Tseng, S. S. Yang, S. T. Wu, C. L. Hsu, and S. J. Chang, "Buffer-facilitated epitaxial growth of ZnO nanowire," Cryst. Growth Des., 5(2), 579-83 (2005).

[34] Y. F. Gao and M. Nagai, "Morphology evolution of ZnO thin films from aqueous solutions and their application to solar cells," Langmuir, 22(8), 3936-40 (2006).

[35] X. D. Wu, L. J. Zheng, and D. Wu, "Fabrication of superhydrophobic surfaces from microstructured ZnO-based surfaces via a wet-chemical route," Langmuir, 21(7), 2665-67 (2005).

[36] R. B. Peterson, C. L. Fields, and B. A. Gregg, "Epitaxial chemical deposition of ZnO nanocolumns from NaOH," Langmuir, 20(12), 5114-18 (2004).

[37] T. Y. Liu, H. C. Liao, C. C. Lin, S. H. Hu, and S. Y. Chen, "Biofunctional ZnO nanorod arrays grown on flexible substrates," Langmuir, 22(13), 5804-09 (2006).

[38] E. Hosono, S. Fujihara, I. Honna, and H. S. Zhou, "The fabrication of an upright-standing zinc oxide nanosheet for use in dye-sensitized solar cells," Adv. Mater., 17(17), 2091-+ (2005).

[39] Y. Masuda, N. Kinoshita, F. Sato, and K. Koumoto, "Site-selective deposition and morphology control of UV- and visible-light-emitting ZnO crystals," Cryst. Growth Des., 6(1), 75-78 (2006).

[40] Y. Masuda and K. Kato, "High c-Axis Oriented Stand-Alone ZnO Self-Assembled Film," Cryst. Growth Des., 8(1), 275-79 (2008).

[41] Y. Masuda and K. Kato, "Self-supported Zn5(CO3)2(OH)6 film formation at air-liquid interface," Trans. Mater. Res. Soc. Jpn, 32(3), 739-42 (2007).

[42] M. D. Gao, M. M. Li, and W. D. Yu, "Flowerlike ZnO nanostructures via hexamethylenetetramine-assisted thermolysis of zinc-ethylenediamine complex," J. Phys. Chem. B, 109(3), 1155-61 (2005).

[43] H. Tokudome, Y. Yamada, S. Sonezaki, H. Ishikawa, M. Bekki, K. Kanehira, and M. Miyauchi, "Photoelectrochemical deoxyribonucleic acid sensing on a nanostructured TiO2 electrode," Appl. Phys. Lett., 87(21), 213901 (2005).

[44] A. H. Nobari and M. Halali, "An investigation on the calcination kinetics of zinc carbonate hydroxide and Calsimin zinc carbonate concentrate," Chemical Engineering Journal, 121(2-3), 79-84 (2006).

[45] D. Andeen, L. Loeffler, N. Padture, and F. F. Lange, "Crystal chemistry of epitaxial ZnO on (111) MgAl2O4 produced by hydrothermal synthesis," Journal of Crystal Growth, 259(1-2), 103-09 (2003).

[46] L. Vayssieres, K. Keis, A. Hagfeldt, and S. E. Lindquist, "Three-dimensional array of highly oriented crystalline ZnO microtubes," Chem. Mater., 13(12), 4395-+ (2001).

[47] Ke Yu, Zhengguo Jin, Xiaoxin Liu, Juan Zhao, and Junyi Feng, "Shape alterations of ZnO nanocrystal arrays fabricated from NH3 · H2O solutions," Applied Surface Science, 253(8), 4072-78 (2007).

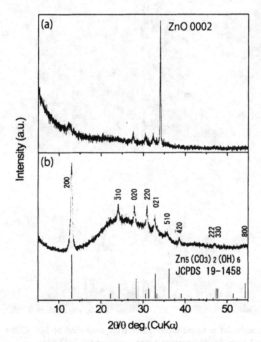

Figure 1. XRD diffraction patterns of (a) a stand-alone ZnO film pasted on a Si wafer and (b) a stand-alone $Zn_5(CO_3)_2(OH)_6$ film pasted on a Si wafer.

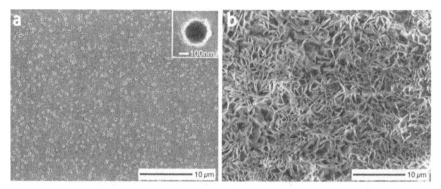

Figure 2. SEM micrographs of a stand-alone ZnO self-assembled film. (a) Air-side surface of a ZnO film. (Insertion) Magnified area of surface hole having hexagonal shape. (b) Liquid-side surface of a ZnO film.

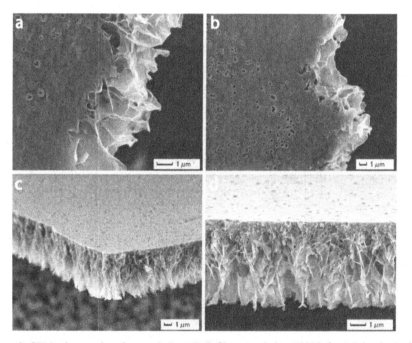

Figure 3. SEM micrographs of a stand-alone ZnO film annealed at 500°C for 1 h in air. (a, b) Edges of a ZnO film. (c, d) Fracture cross section of a ZnO film.

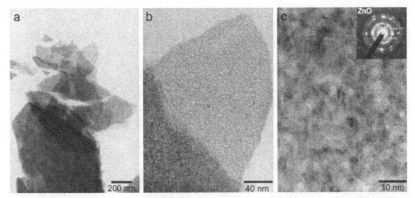

Figure 4. (a) TEM micrograph of ZnO nano-sheets. (b, c) Magnified area of (a). (Insertion) Electron diffraction pattern of ZnO.

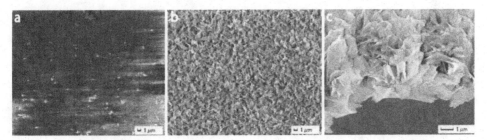

Figure 5. SEM micrographs of a stand-alone $Zn_5(CO_3)_2(OH)_6$ film. (a) Air-side surface of a $Zn_5(CO_3)_2(OH)_6$ film. (b) Liquid-side surface of a $Zn_5(CO_3)_2(OH)_6$ film. (c) Edges of a $Zn_5(CO_3)_2(OH)_6$ films.

FABRICATION OF THE FINESTRUCTURED ALUMINA POROUS MATERIALS WITH NANOIMPRINT METHOD

Hong Dae KIM, Tadachika NAKAYAMA[†], Jun YOSHIMURA, Kazuyoshi IMAKI, Takeshi YOSHIMURA[*], Hisayuki SUEMATSU, Tsuneo SUZUKI and Koichi NIIHARA

Extreme Energy-Density Research Institute, Nagaoka university of technology, 1603-1 Kamitomioka, Nagaoka 940-2188
*Department of Physics and Electronics, Graduate School of Engineering, Osaka Prefecture University, 1-1 Gakuencho, Nakaku, Sakai 599-8531

ABSTRACT

Nanoimprint lithography (NIL), is recognized as one of the candidates for next generation nanolithography. Usually nanoimprint equipment is highly expensive because the nano-scale mold needs to be hardened with thermal and photonic treatment after casting. In this study, we attempted to fabricate finestructured alumina porous materials without using highly expensive equipment. This method is based on the idea that poly vinyl alcohol (PVA) is simply detached from silicon mold by peeling. The experiment process used PVA and alumina nano-size particle, mixed in water solution. The mixed solution was then put into a micro-scale Si-mold and hardened at room temperature. After hardening the PVA and alumina nano-size particle mixed solution was detached from the micro-scale Si-mold. Throughout burn-out and low-temperature sintering, we confirmed the fabrication of finestructured micrometer-size alumina with nano-size pores.

INTRODUCTION

The ability to fabricate structures from the micro- to the nanoscale with high precision in a wide variety of materials is of crucial importance to the advancement of micro- and nanotechnology and the nanosciences. Nanoimprint lithography (NIL), which is recognized as one of the candidates for next generation nanolithography, was first reported by Chou. NIL can not only create resist patterns, as in lithography, but can also imprint functional device structures is various polymers, which can lead to a wide range of applications in electronics, photonics, data storage, and biotechnology. [1-3] However NIL includes expensive and complex processes. Usually NIL equipment is highly expensive because the nano-scale mold needs hardening with thermal and photonic treatments after casting. Recently, several polymer materials have been applied in most types of usual NIL as the template materials, including polydimethyl siloxane (PDMS), polymetyl methacrylate (PMMA), polyurethane (PU), and polyvinylalcohol (PVA). [4] Particularly, the nanoimprint application using PVA polymer materials with following advantages has been numerous concentrations. First of all, PVA has been proposed as a nanoimprint template for a high resolution and low cost, a high Young's modulus, due to its solubility in water. [5] For instance, a conventional lift-off process using polymetyl methacrylate (PMMA) uses acetone as solvent, while a lift-off process using PVA uses water as a solvent. Also PVA has desirable properties for printing, including a high Young's modulus of 1.9 GPa, compared to 1.8 MPa for polydimethylsiloxane (PDMS), which is important for minimizing distortions. [6-10]

In this study, we attempted to fabricate finestructured alumina porous materials, which is one of ceramic materials showing pronounced chemical stability and the outstanding thermomechanical properties in comparison to plastics and metals, without using highly expensive nanoimprint equipment. We used a simple method by combining the advantages of usual nanoimprint method and PVA polymer material. This method is based on the idea that PVA is simply detached from silicon mold by peeling.

EXPERIMENTAL

The principle of fabrication is very simple. Fig. 1. shows the schematic process of this experiment. Si mold (25mm X 25mmX 1mm) was chosen from a commercial product (Kyodo International, Co. Ltd., Japan), which was fabricated by conventional electron-beam lithography and by a dry etching process. Si mold have patterns such as space & line, dot, hole and pattern size ranging from 0.5 μm to 50 μm. PVA (Average MW=500, Wako Pure Chemical Industries, Co. Ltd., Japan) and alumina nano particles (TM-300D, γ-alumina, 10 nm; Taimei Kagaku, Co. Ltd., Japan) were used for raw materials.

Firstly, PVA and alumina nano particles were mixed in water solution and ball-milled with zirconia ball (about 5 mm, 12 h). The surface of Si mold was coated with a fluorine-contained release agent (HD-1100, Harces fluorine chemical, Japan) to detach the PVA /alumina complex from Si-mold. Then, the PVA/alumina complex was put into the patterned silicon mold and dried. The dry process was carried out by temperature/humidity test chamber (KCL-2000, EYELA Co. Ltd., Japan) to minimize distortion of PVA/alumina complex film in condition of temperature (80 °C) and humidity (60%). The dried PVA/Alumina complex was detached from the silicon mold by peeling. The patterned complex was sintered at various sintering temperatures ranging from 1,000 to 1,500 °C. Usually, the PVA burns out during sintering around 500 °C.

Size and surface of alumina patterns was observed using scanning electron microscope (SEM, JSM-6700F, JEOL, Japan) Phase identification was carried out by X-ray diffraction (XRD, RINT 2500, Rigaku, Japan) analysis.

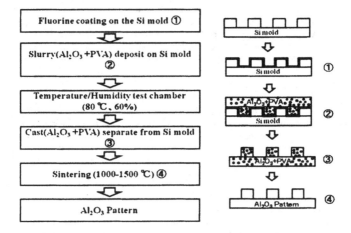

Fig. 1. A schematic drawing of the experimental procedure
for nanoimprint process of alumina sintered body.

RESULTS AND DISCUSSION

It is observed that using various Si mold patterns, such as; dot, hole, line & space types, and pattern sizes ranging from 0.5 μm to 50 μm, imprinted porous alumina patterns, in accordance with those of Si molds, were fabricated. Followed by sintering, the most porous alumina patterns were shrunk in size (SEM image in Fig. 2.). In the case of Si mold pattern having 0.5 μm in size, sub-micron sized porous alumina pattern, down-to about 0.3 μm in size, was formed.

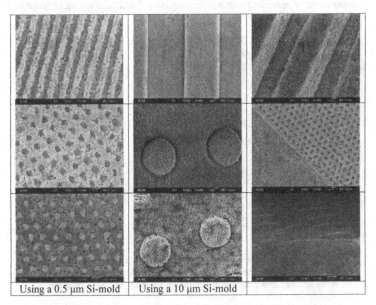

Using a 0.5 μm Si-mold | Using a 10 μm Si-mold

Fig. 2. SEM micrographs of the finestructured alumina sintered body
with nanoimprint processing with several kinds of patterning size and shape.

To optimize the fabrication method of the finestructured alumina porous materials, the amounts of PVA, alumina and water were changed as showed just as Table 1. In case, PVA amounts exceed 5 wt%, PVA starts to educe. While the amount of PVA goes down lower than 2 wt%, the alumina pattern was not formed owing to the decrease of binding strength. When the amount of alumina powder was less than 6 wt% or more than 12 wt% of total amount, the patterns were not fabricated. The optimized alumina patterns were fabricated with 5 wt% of PVA and 6-12 wt% of alumina powder.

Table 1. Improvement Percentages of the Chemical Composition of the Alumina Slurry

	PVA	Al_2O_3	H_2O	Remark
1	2wt%	6wt%	92wt%	
2	2wt%	12wt%	86wt%	
3	2wt%	6wt%	92wt%	
4	2wt%	12wt%	86wt%	
5	**5wt%**	**6wt%**	**89wt%**	**Formation**
6	**5wt%**	**12wt%**	**83wt%**	**of film**
7	5wt%	18wt%	77wt%	

Fig. 3. shows surface SEM images of alumina pattern after sintering. PVA is removed by evaporation from around 500 °C resulting in the formation of pores. The pore size and porosity is dependent on the amount of PVA present and sintering temperature. The porosity and surface area of the various patterned alumina could be controlled with a processing control such as sintering temperature (1000 °C - 1500 °C).

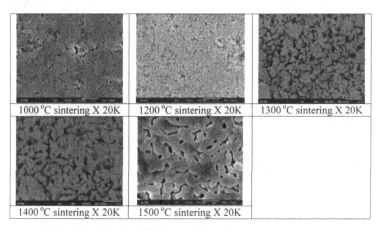

1000 °C sintering X 20K	1200 °C sintering X 20K	1300 °C sintering X 20K
1400 °C sintering X 20K	1500 °C sintering X 20K	

Fig. 3. High magnification SEM images of the alumina pattern after sintering.
After sintered under the high temperature, grain growth was confirmed.
After sintered under the low temperature, the fine pore was observed.

In case of alumina nano powder, phase transition of alpha phase from gamma phase was confirmed with XRD patterns of alumina. After sintering, the pattern shrinks down to 16~31% (Fig. 4.). Shrinking rate of sintered alumina patterns was increased according to the elevation of sintering temperature. For this shrinking mechanism, the use of micron size Si mold can lead a sub-micron alumina patterns after the occurrence of shrinking.

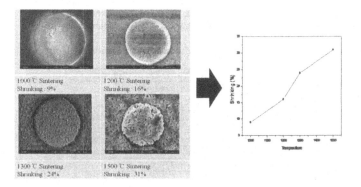

Fig. 4. Shrinking rate of the sintered alumina with fine structure according to the sintering temperature.

CONCLUSIONS

We confirmed the fabrication of finestructured alumina porous materials with nano-size pores using a simple process. Porosity and shrinkage of alumina patterns could be determined by controlling the sintering temperature, and the shrinkage control facilitated the fabrication of sub-micron alumina pattern in spite of the use of micron size Si mold. In addition, surface area of the patterned alumina materials could be controlled by simple sintering temperature control. We believe that the proposed process to fabricate ceramic finestructured patterning has merits in the simple process and low production cost, therefore it can be applied in the various field. It is considered that the submicron-patterned alumina with nanoporous structure is likely to be widely applied on not only catalyst support but also super water repellent and heat-proof material because of high surface area.

ACKNOWLEDGMENT

This study was financially supported by supported by Strategic Information and Communications R&D Promotion Program (SCOPE), Ministry of Internal Affairs and Communications, Japan and Knowledge Cluster Project, Ministry of Education, Culture, Sports, Science and Technology, Japan.

REFERENCES

[1] S. Y. Chou, P. R. Krauss, P. J. Renstrom, Imprint of sub-25 nm vias and trenches in polymers, *Appl. Phys. Lett.*, **67**, 3114-3116 (1995).
[2] Haisma, J., Verheijen, M. and Heuvel, K, Mold-assisted nanolithography: A process for reliable pattern replication, *J. Vac Sci. Technol. B.*, **14**[6], 4124-4128 (1996).
[3] L.jay Guo, Nanoimprint Lithography: Methods and Material Requirements, *Adv. Mater.*, **19**, 495-513 (2007).
[4] Ki-don Kim, Jun-ho Jeong, Altun Ali, Dong-il Lee, Dae-geum Choi, Eung-sug Lee, Replication of an UV-NIL stamp using DLC coating, *Microelectronic Engineering.*, **84**, 899-903 (2007).
[5] C. D. Schaper, Patterned Transfer of Metallic Thin Film Nanostructures by Water-Soluble Polymer Templates, *NanoLetters.*, **3**[9], 1305-1309 (2003).
[6] C. D. Schaper, Alan Miahmahri, Polyvinyl alcohol templates for low cost, high resolution, complex printing, *J. Vac. Sci. Technol*, B., **22**[6], 3323-3326 (2004).
[7] Ken-ichiro Nakamatsu, Katsuhiko Tone, Shinji Matsui, Nanoimprint and Lift-Off Process Using Poly(vinyl alcohol), *J. Appl. Phys.*, **44**[11]. 8186-8188 (2005).
[8] Yoshihiko Hirai, Takashi Yoshikawa, Masatoshi Morimatsu, Masaki Nakajima, Hiroaki Kawata, Fine pattern transfer by nanocasting lithography, *Microelectronic Engineering*, **78-79**, 641-646 (2006).
[9] H.-C. Scheer, N. Bogdanski, M. Wissen, T. Konishi, Y. Hirai, Profile evolution during thermal nanoimprint, *Microelectronic Engineering*, **83**[4-9], 843-846, (2006).
[10] Kenji Sogo, Masaki Nakajima, Hiroaki Kawata, Yoshihiko Hirai, Reproduction of fine structures by nanocasting lithography, *Microelectronic Engineering*, **84**[5-8], 909-911 (2007).

STRUCTURE CONTROL OF THE NANOTUBE/NANOPARTICLE HYBRID MATERIALS WITH SONOCHEMICAL PROCESSING

Masahiro Terauchi, Tadachika Nakayama, Hisayuki Suematsu, Tsuneo Suzuki and Koichi Niihara

Extreme Energy-Density Research Institute, Nagaoka University of Technology, 1603-1 Kamitomioka-cho, Nagaoka 940-2188, Japan

ABSTRACT

TiO_2 nanotube was filled with Ag nanoparticles by sonochemical process in methanol solution. Ag nanoparticles confirmed within TiO_2 nanotube was about 5 nmφ in diameter, and had elliptical shape such seem occupy the nanotube. According to thermal analysis of composited TiO_2 nanotube /Ag nanoparticle nanosystem, peculiar second phase weight loss, exothermic heat, and generation of CO_2 were confirmed at about 250 °C. These were suggested that methanol or other solution derived by its pyrolysis or oxidation were discharged, and burned by catalysis of Ag nanoparticles which been around TiO_2 nanotube.

INTRODUCTION

Titania nanotube (TiO_2NT) possesses a lot of absorbing characteristic such as high specific surface area[1,2], ion-changeable ability[3], photocatalytic ability[4], and gas-adsorption properties[5] therefore it had been investigated for extensive applications. So far, as synthetic method of TiO_2NT, template method[6-8], sol-gel method[1], anodic oxidation method[9-12], and hydrothamal method[4, 13-18] had been confirmed. In these facture, the hydrothermal method confirmed by Kasuga $et.al$[13] can fabricate TiO_2NT with average inner diameter of about 5 nmφ, outer diameter of about 10 nmφ, and length of over 100 nm similar to carbon nanotube (CNT). Although, these TiO_2NT all has open end structure different from in case of the CNT that has not only open end structure but also the closed end structure.

Previously, CNT had been much investigated due to its peculiar physical and electronic property. CNT is also studied as molecular transporters and drug delivery system with this unique structure[19-21]. However, in this case, it is necessary to modify outer wall in functional group or high molecular as the post-treatment. In contrast, it is known that the surface in the TiO_2NT being covered with many -OH bases and easily make chemisorption. Moreover, the heat-resistance of TiO_2NT is low, since the TiO_2NT is the metastable structure. The structure collapses on usual TiO_2NT at about 350 °C, and it begins to change in the aggregate of particle[22]. Due to these characteristic, TiO_2NT also can be expected to apply to fluidic industry and liquid-ejection system in nano area.

However, because of it unique structure, to fabricate molecular transporter with TiO_2NT, the nanotube have to be closed by other elements like metal or ceramics nano particles. Previously some researchers have reported about TiO_2NT containing nano particle of some metals, which were synthesized by CVD process[23]. Although, the TiO_2NT has to be treated under the high temperature to decompose the organometallic complex. The liquid inside and outside of the TiO_2NT will vaporize while these kinds of high temperature process. Therefore, other process which can fabricate the TiO_2NT containing nano particle in low temperature is demanded.

Recently, sonochemical process, which is one of the low temperature processes to fabricate nano particle are reported[24, 25]. Sonochemical process is a synthesis method that applies cavitation and following hot-spot resulted by ultrasonic irradiation. It is known that cavitasion and hot-spot beget several effect, such as synthesis of nanoparticles resulted by thermal decomposition of metal complex or metal compound, decomposition of macro molecular, and development of other reaction path by generation of radicals[24, 25].

Therefore in this study, composite of TiO_2NT filled with solution and nanoparticles was attempted by using sonochemical process. Furthermore, capability of the TiO_2NT such as system of active ejection by heat-treatment were presented.

EXPERIMENTAL PROCEDURE

TiO_2NT was synthesized by hydrothermal method and ion-exchange by acid treatment from anatase-type TiO_2 (mean particle size is 200 nm, Wako Chem. Co. Ltd., Japan). One gram of anatase-type TiO_2 and 40 ml of 8 M NaOH solution were poured into a bonbe, and treated at 130 °C for 72 h. Obtained product were washed with 0.5 mol/L HCl solution and ion-exchanged water. Finally, product were dried by suction filtration.

Twenty milligram of dried products were mixed with 20 mg of Ag_2O (mean particle size is 5μm, Kojundo Chemical Co. Ltd., Japan) and 50 ml of distilled water or methanol (>99.8 %, Wako Chemical Co. Ltd., Japan). Sonochemical process with frequency of 200 kHz, output of 430 W, and irradiation time of 150 min. was carried out for this slurry at solution temperature of 30 °C.

For obtained slurry, nanostructural observation was carried out by transmission electron microscopy (TEM) (JEOL JEM-2000FX) with an operating voltage of 200 kV. The phase identification was carried out by X-ray diffraction (XRD) (RINT-2500PC, Rigaku Co. Ltd. Japan) technique using Cu-$K_{\square}$ and selected area electron diffraction (SAD). XRD patterns of products were obtained in 2θ between 10° to 120° with a step of 0.02° and a scan speed of 2° minute^{-1}. The thermal analysis was carried out by TG-DTA (Bruker AXS TG-DTA2000SA) and TG-MASS (Bruker AXS MS9610) with rate of temperature increase of 5°C /min. and in He gas flow.

The sample for TEM was obtained by putting a drop to the Cu micro-grid and drying in the atmosphere at room temperature. The measurement sample for XRD was prepared by putting a few drop to Si wafer and drying in the atmosphere at room temperature, therefore XRD pattern was measured quickly after sonochemical process. Thermal analysis was carried out about dried products at 80 °C in air.

RESULTS AND DISCUSSION

Figure 1. shows the appearance of the slurry, (a) before and (b) after ultrasonic irradiation. In spite of the slurry which is before irradiating ultrasonic was precipitated, the slurry after irradiation was dispersed well. Furthermore, the color of the slurry changed itself from white to black.

Figure 1 The appearance of the sample: (a) before the sonochemical process (mixture of the TiO_2NT, Ag_2O and methanol) and (b) after the sonochemical processing. Ag_2O was reduced by sonochemical processing to the Ag nanoparticle and well dispersed.

Figure 2. shows the XRD patterns of product after ultrasonic irradiation (Sample in Figure 2.). XRD result of as-synthesized TiO_2NT sample suggested that anatase phase and TiO_2NT $(H_2Ti_nO_{2n+1})^{3,14}$ phase were mixed. Then, from XRD result of product after ultrasonic irradiation,

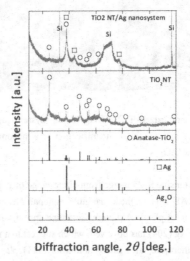

Figure 2 XRD pattern of as-synthesized TiO_2NT and $TiO_2NT/AgNP$ nanosystem fabricated by sonochemical processing.

diffraction peak from anatase phase and silver phase were ascertained. In addition, because of diffraction peaks from Ag_2O were not found, it was suggested that Ag_2O was deoxidised by thermal decomposition or enhanced chemical reaction as follows caused by the cavitation.

$$Ag_2O + CH_3OH \rightarrow 2Ag + HCHO + H_2O$$

Figure 3 shows the TEM blight field images (BFI) of products by sonochemical process for 150 minutes. The nanoparticle of about 10 nmφ in diameter were on outer wall of TiO_2NT. Furthermore, such nanoparticle of about 5 nmφ in diameter were confirmed also within TiO_2NT. Additionally those nanoparticles entrapped within TiO_2NT have elliptical shape such seem occupy the nanotube. Therefore, sonochemical process for composition of TiO_2NT including particles has been successful.

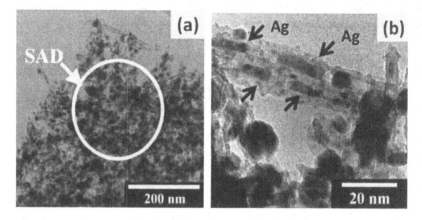

Figure 3 TEM images of synthesized TiO_2NT/AgNP nanosystem (a) low magnification and SAD analysis area (white circle) and (b) high magnification. Ag nanoparticles (black arrowed) were encapsulated the inside of the TiO_2 NT.

Figure 4 shows extensive SAD image from circle range of Figure 3. According to calculated lattice spacing from this SAD image, those nanotube/nanoparticle nanosystem were composed of anatase-type TiO_2 phase and Ag phase. Although the lattice spacing of these phases are too close to each other, diffraction from Ag (204) has observed, which suggest that the nanoparticles that observed in Figure 3 was Ag. Furthermore, also diffraction from (111) plane of raw Ag_2O could not be observed, so that it was supported that nanoparticles on TiO_2NT were Ag metal nanoparticle.

As shown in Figure 3 (b), nanoparticles entrapped within TiO_2NT have elliptical shape such seem occupy the nanotube. That suggests grain growth was continued inside of the tube. Therefore, such Ag containing process is incomplete without through monatomic or ionic state of Ag. On the other hand, TiO$_2$NT and Ag particle are to disperse in methanol usually because of zeta-potential of both are negative.   Accordingly, it is suggested that Ag nanoparticle occupying TiO$_2$NT are result of filling up by methanol with Ag$^+$ and result of subsequent precipitation of Ag particles onto inner wall. In this case, it is supposed that methanol, other hydrocarbon (such as formic aldehyde), and water are satisfied in the cavity of among two Ag nanoparticle.

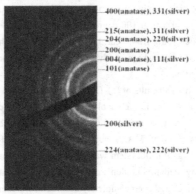

400(anatase), 331(silver)

215(anatase), 311(silver)
204(anatase), 220(silver)
200(anatase)
004(anatase), 111(silver)
101(anatase)

200(silver)

224(anatase), 222(silver)

Figure 4 SAD pattern of the TiO$_2$NT/AgNP nanosystem (analysis area within white circle in Fig. 3 (a)).

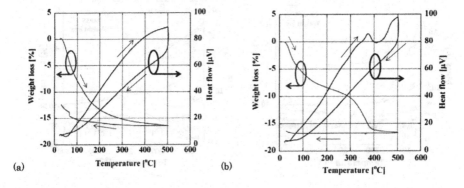

(a)

(b)

Figure 5 TG-DTA results of (a) as-synthesized TiO$_2$NT and (b) TiO$_2$NT/AgNP nanosystem fabricated by the sonochemical process in distilled water.

Figure 5 shows the TG-DTA results of (a) as-synthesized TiO_2NT and (b) $TiO_2NT/AgNP$ nanosystem fabricated by the sonochemical process in distilled water. There is the large amount of OH base on the sureface of the titania nanotube. Large amount of water is adsorbed on the surface of the titania nanotube[22]. In figure 5(a), the weight loss of ca.15% was observed by heating the titania nanotube to 500 degrees. This means that the adsorption water evaporates from not only external wall in the titania nanotube but also internal wall. In fugure 5(b), the elimination of adsorption water becomes a two stage in titania nanotube which was combined with the silver nanoparticle. This originates for the silver nanoparticle making the cup to be the titania nanotube. The adsorption water of external wall in the nanotube desorbs near 100 degrees, and it is indicated that the adsorption water of the inside in the nanotube is desorbing from near 350 degrees in which the structure of the nanotube deformed. Titania nanotube composite which involved the liquid except for the water was synthesized, and these phenomena were verified.

Figure 6 shows TG-MASS results of (a) TiO_2NT powder and (b) $TiO_2NT/AgNP$ nanosystem fabricated by the sonochemical process in methanol. By causing ultrasonic wave chemical reaction in the methanol, it was considered that the methanol could be involved in the titania nanotube. For the TiO_2NT powder, any peak signals of MASS number have detected for this temperature range. Moreover, in spite of $TiO_2NT/AgNP$ nanosystem was synthesized in methanol, peak signals of M=29, 31, 32, which should appear by methanol couldn't detect. This fact indicates that the methanol which was adsorbed on outer wall surface had been completely desorbed before analysis by preliminary

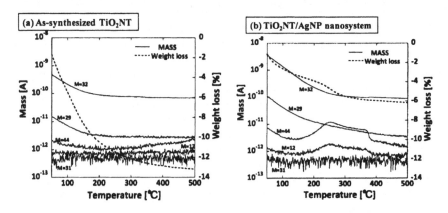

Figure 6 TG-MASS results of (a) as-synthesized TiO_2NT and (b) $TiO_2NT/AgNP$ nanosystem fabricated in methanol. The MASS No.44 gas (CO_2 gas) derived the methanol was detected from 250 to 380 °C. This temperature is almost same as TiO_2NT structure change temperature.

drying. However, TiO_2NT/Ag powder showed peak signals of M=12, 44 at about 250 °C, which MASS number typically show in generation of CO_2. Taking on account that there were drastic weight loss at that temperature, it is evident that CO_2 was generated by combustion. Therefore, these fact suggest that some kind of organic solution like methanol should be filled inside TiO_2NT and released by the heat treatment. Furthermore, the released solutions should be burned by catalysis of Ag nanoparticles which is modifying around TiO_2NT.

CONCLUSION

We have successfully synthesized the TiO_2NT which is filled with solution and nanoparticles by irradiating ultrasonic to the mixed powder of TiO_2NT and Ag_2O in methanol and its thermal property was observed. From TEM analysis, synthesized sample was composition of TiO_2NT and nanopartices of Ag. Furthermore, a few nm size particles were confirmed inside TiO_2NT. From TG-MASS analysis, synthesized TiO_2NT losses its weight drastically at around 250 °C and emits CO_2 gas, which could be caused by emitting the solution from TiO_2 nano tube by reformation its structure from TiO_2NT to nanorod.

ACKNOWLEDGMENT

This study was financially supported by supported by Strategic Information and Communications R&D Promotion Program (SCOPE), Ministry of Internal Affairs and Communications, Japan and Knowledge Cluster Project, Ministry of Education, Culture, Sports, Science and Technology, Japan.

REFERENCES

[1]T. Kasuga, M. Hiramatsu, A. Hoson, T. Sekino, K. Niihara, Formation of Titanium Oxide Nanotube, *Langmuir*, 14, 3160–3163 (1998).

[2]S. Zhang, J. Zhou, Z. Zhang, Z. Du, A.V. Vorontsov, Z. Jin, Morphological structure and physicochemical properties of nanotube TiO_2, *Chin. Sci. Bull.,.* 45(16), 1533-1536 (2000).

[3]X. Sun, Y. Li, Synthesis and Characterization of Ion-Exchangeable Titanate Nanotubes, *Chem. Eur. J.*, 9, 2229-2238 (2003).

[4]M. Zhang, Z. Jin, J. Zhang, X. Guo, J. Yang, W. Li, X. Wang, Z. Zhang, Effect of annealing temperature on morphology, structure and photocatalytic behavior of nanotubed $H_2Ti_2O_4(OH)_2$, *J. Mol. Catal.*, A 217, 203–210 (2004).

[5]P. Umek, P. Cevc, A. Jesih, A. Gloter, C.P. Ewels, D. Arcon, Impact of Structure and Morphology on Gas Adsorption of Titanate-Based Nanotubes and Nanoribbons, *Chem. Mater.*, 17, 5945-5950 (2005).

[6]P. Hoyer, Formation of a Titanium Dioxide Nanotube Array, *Langmuir*, 12, 1411–1413 (1996).

[7]J.H. Jung, H. Kobayashi, K.J.C. van Bommel, S. Shinkai, T. Shimizu, Creation of Novel Helical Ribbon and Double-Layered Nanotube TiO_2 Structures Using an organogel Template, *Chem. Mater.*,

14, 1445–1447 (2002).

[8]J.H. Lee, I.C. Leu, M.C. Hsu, Y.W. Chung, M.H. Hon, Fabrication of aligned TiO_2 one-dimensional nanostructured arrays using a one-step templating, *J. Phys. Chem. B*, **109**, 13056–13059 (2005).

[9]D. Gong, C.A. Grimes, O.K. Varghese, W. Hu, R.S. Singh, Z. Chen, E.C. Dickey, Growth of nano-scale hydroxyapatite using chemically treated titanium oxide nanotubes, *J. Mater. Res.*, **16**, 3331–3334 (2001).

[10]O.K. Varghese, D. Gong, M. Paulose, C.A. Grimes, E.C. Dickey, Crystallization and high-temperature structural stability of titanium oxide nanotube arrays, *J. Mater. Res.*, **18** 156–165 (2003).

[11] A. Ghicov, H. Tsuchiya, J.M. Macak, P. Schmuki, Titanium oxide nanotubes prepared in phosphate electrolytes, *Electrochem. Commun.*, 7, 505–509 (2005).

[12] H. Tsuchiya, J.M. Macak, L. Taveira, E. Balaur, A. Ghicov, K. Sirotna, P. Schmuki, Self-organized TiO2 nanotubes prepared in ammonium fluoride containing acetic acid electrolytes, *Electrochem. Commun.*, 7, 576–580 (2005).

[13] T. Kasuga, M. Hiramatsu, A. Hoson, T. Sekino, K. Niihara, Titania Nanotubes Prepared by Chemical Processing, *Adv. Mater.*, 11 1307–1311 (1999).

[14]G.H. Du, Q. Chen, R.C. Che, Z.Y. Yuan, L.M. Peng, The structure of trititanate nanotubes, *Appl. Phys. Lett.*, 79, 3702–3704 (2001).

[15] Q. Chen, W.Z. Zhou, G.H. Du, L.M. Peng, Trititanate Nanotubes Made via a Single Alkali Treatment, *Adv. Mater.*, 14, 1208–1211 (2002).

[16]Q. Chen, G.H. Du, S. Zhang, L.M. Peng, The structure of trititanate nanotubes, *Acta Cryst.*, **B 58** , 587–593 (2002).

[17]S. Zhang, L.M. Peng, Q. Chen, G.H. Du, G. Dawson, W.Z. Zhou, Formation mechanism of $H_2Ti_3O_7$ nanotubes, *Phys. Rev. Lett.*, 91, 256103-1:4 (2003).

[18]J. Yang, Z. Jin, X. Wang, W. Li, J. Zhang, S. Zhang, X. Guo, Z. Zhang, Study on composition, structure and formation process of nanotube $Na_2Ti_2O_4(OH)_2$, *Dalton Trans.*, 3898–3901 (2003).

[19]Y. Gogotsi, N. Naguib, J.A. Libera, In situ chemical experiments in carbon nanotubes, *Chem. Phys. Lett.*, **365** 354–360 (2002).

[20]A. Bianco, M. Prato, Can Carbon Nanotubes be Considered Useful Tools for Biological Applications?, *Adv. Mater.*, 15(20), 1765–1768 (2003).

[21] N.W.S. Kam, T.C. Jessop, P.A. Wender, H. Dai, Nanotube Molecular Transporters: Internalization of Carbon Nanotube Protein Conjugates into Mammalian Cells, *J. Am. Chem. Soci.*, 126(22), 6850-6851 (2004).

[22] R. Yoshida, Y. Suzuki, S. Yoshikawa, Effects of synthetic conditions and heat-treatment on the structure of partially ion-exchanged titania, *Mater. Chem. Phys.*, 91, 409–416 (2005).

[23]A. Nakahira, T. Kubo, Y. Yamasaki, T. Suzuki, Y. Ikuhara, Synthesis of Pt-Entrapped Titanate Nanotubes, *Jpn. J. Appl. Phys.*, 44(22), L690-L692 (2005).

[24]K.S. Suslick, D.A. Hammerton, R.E.Cline, Sonochemical hot spot, *J. Am. Chem. Soc.*, **108**, 5641 (1986).

[25]Y. Hayashi, H. Takizawa, M. Inoue, K. Niihara, K. Suganuma, Various Applications of Silver Nano-Particles By Ultrasonic Eco-Fabrication, *IEEE Trans. on Electronics Packaging Manufacturing*, **28**, 338 (2005).

EFFICIENT PHOTOCATALYTIC DEGRADATION OF METHYLENE BLUE WITH CuO LOADED NANOCRYSTALLINE TiO$_2$

Arun Kumar Menon and Samar Jyoti Kalita*
Department of Mechanical, Materials and Aerospace Engineering
University of Central Florida
P.O. Box 162450
Orlando, FL 32816-2450

*Current Affiliation: Engineered Surfaces Center, School of Engineering and Mines
University of North Dakota, Grand Forks, ND 58202-8391

ABSTRACT

Nanocrystalline titanium dioxide (TiO$_2$, anatase) is a promising material in solving various environmental problems because of its unique photocatalytic activity and high photoelectric performance. However, the photocatalytic activity of phase pure nanocrystalline anatase is low because of easy recombination of photoinduced holes and electrons. This research enhanced the photocatalytic efficiency of nanocrystalline anatase by loading with copper oxide (CuO). In this effort, a series of CuO loaded TiO$_2$ photocatalysts were prepared and analyzed. The photodegradation of methylene blue solution in the presence of these photocatalysts under UV radiation was studied. The color of the methylene blue solution faded gradually with time indicating photocatalysis. Results showed that the photocatalytic activity of nanocrystalline anatase was enhanced by the addition of CuO till 20 wt% addition, after that the photocatalytic activity notably dropped. XPS results showed that the deposited copper formed Ti-O-Cu bond on the surface of TiO$_2$.

INTRODUCTION

Titanium dioxide is a very commonly used photocatalytic material as it possesses unique characteristics such as easy handling, low cost, low toxicity, chemically stable [1] and is also robust under UV illumination. These advantages make TiO$_2$ a useful material in environmental purification [2] , gas sensors [3], paints, pigments [4], air-cleaning devices, removal of NO$_x$ and SO$_x$ in controlling pollution, etc. TiO$_2$ generally exhibits three different polymorphs namely anatase, rutile and brookite. The anatase and rutile phases have tetragonal symmetry—anatase has body centered tetragonal structure, rutile possesses simple tetragonal structure whereas brookite possesses rhombohedral structure. In both structures, slightly distorted octahedra are the basic building blocks, which consist of a titanium atom surrounded by six oxygen atoms in a more or less distorted octahedral configuration. Anatase belongs to D_{4h}^{14}-P42 / mnm space group (lattice constant a = 0.4584 nm, c = 0.2953nm, c/a = 0.664), while rutile belongs to D_{4h}^{19}- I41 / amd space group (lattice constant a = 0.3733 nm, c = 0.937 nm, c/a = 2.51) [5]. Additional phases of TiO$_2$ are also known to exist. Amongst, antase phase exhibits the best photocatalytic activity [6]. The photocatalytic activity of TiO$_2$ depends on several factors including the crystallite size, crystallinity of the phase and specific surface area. Photocatalytic activity of TiO$_2$ is high for powder with high surface area because of increased active sites, which improves absorbance capability on surface.

TiO$_2$ photocatalyst can only be excited by near-UV light. The photocatalytic activity of TiO$_2$ is due to photo-induced electrons and the corresponding positive holes that are formed. These species are responsible for initiating the photcatalytic reactions. Furthermore, it is believed that the photocatalytic activity of TiO$_2$ is low, because the generated photoinduced holes easily recombine with photoinduced electron thus decreasing the photocatalytic efficiency [7]. If the recombination rate is very fast, there is not enough time for photoreaction, which results in little or no photocatalytic effect. The energy gap of pure anatase phase is large (3.2 eV) and there is a need to narrow this band gap energy by loading TiO$_2$ with other metal ions. Several efforts were made by various researchers to improve the photocatalytic efficiency of TiO$_2$ through metal ion loading. A number of metals including Pd, Pt, Ni [8-10], Cu [11], Zn [12], Cr [13], Ce [14], Bi [15], Fe and Ho [16], Ag [17] have been investigated. The deposited metal ions on TiO$_2$ surface trap and capture the photoinduced electrons or holes, leading to reduction of electron-hole recombination, thereby improving the photocatalytic efficiency [18]. The photocatalysts for bio-degradation can be processed by various techniques. One of the important ones includes the photoreduction technique. This technique involves chemical reduction reactions induced by a source of radiant energy. Y. Xu *et al.* [11] demonstrated that 0.16 mol% Cu$_2$O loaded TiO$_2$ prepared by photoreduction technique effectively degrades methylene blue solution under UV source.

Here, the photocatalytic activities of TiO$_2$ and CuO loaded TiO$_2$ were evaluated by studying the degradation of aqueous methylene blue solution under the action of UV light. The nanocrystalline TiO$_2$ powder for this study was synthesized by a simple Sol-gel technique and the photocatalyst was prepared by UV irradiation for 1 h.

EXPERIMENTAL PROCEDURE

Synthesis of nanocrystalline TiO$_2$ powder through Sol-Gel processing
Nanocrystalline titanium dioxide powder with an average crystallite size of 5-15 nm was synthesized using sol-gel technique [19]. The flowchart showing the steps involved in the synthesis of nanocrystalline TiO$_2$ powder is as shown in Figure 1(a). The hydrolysis reaction of Titanium tetraisopropoxide (TTIP) leads to the formation of TiO$_2$, which is represented by the following reaction:

$$TTIP + 2H_2O \rightarrow TiO_2 + 2C_3H_7OH \qquad (1)$$

The sol-gel process offers unique advantages such as better control over stoichiometric composition, ease of synthesis, better homogeneity and production of high purity powder. Processing conditions, such as chemical concentration, the pH, peptization time, calcinations time and temperature have a great influence on the particle size and phase purity of the final powder.

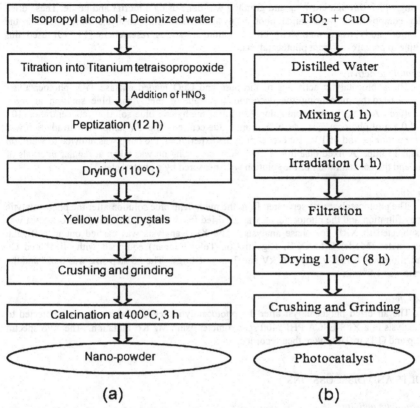

Figure 1. Schematics of the process used in preparation of nanocrystalline anatase powder (a), and CuO loaded anatase photocatalysts

Preparation of photocatalyst

A series of CuO/TiO₂ photocatalysts loaded with 10-40 wt% of CuO were prepared by the photoreduction method. This technique has been used by others to prepare Cu₂O-TiO₂ photocatalyst [11]. We adapted and optimized this process to develop CuO loaded TiO2 photocatalyst. It has been shown that copper when deposited on titanium dioxide leads to its recovery in photoreductive [20] or photosynthetic forms [21]. This was the basis of our selection of photoreduction process in this study. The steps involved in this procedure are as shown in Figure 1(b). Initially, TiO₂ was mixed with differing amounts of CuO (10-40 wt %). Each composition was separately mixed in distilled water using a high-speed magnetic stirrer for 1 h. These mixtures were then irradiated with a 600W UV lamp (Newport Oriel Instruments, Model# 69920) which generates light in 600-100 nm range. The irradiation time was 1 h. The final solution was filtered to collect the solid residue. Residue was dried in an oven for 8 h at 110°C, following which through crushing and grinding was done using a mortar and pestle. These dried powder compositions were called photocatalysts. These photocatalysts were used for the degradation study using methylene blue solution. Various researchers [11, 13] used this technique to prepare different photocatalysts.

Photocatalytic activity

The photocatalytic activities of the pure and CuO loaded anatase TiO$_2$ photocatalysts were evaluated by measuring the decomposition rate of methylene blue solution at room temperature. A glass beaker containing 100 ml of methylene blue solution (concentration 10^{-3} M) and 0.5 g of photocatalyst was placed under the centre of the UV lamp. The irradiation was done for a time period of 1 h. Before starting the experiment, the lamp was allowed to warm up for about 15 min. A fixed quantity of methylene blue solution was taken at regular intervals of 10 min and the concentration of the solution was measured by a photometer.

Phase Analysis

The photocatalyst was separated from the methylene blue solution after the photocatalytic study by filtration and the photocatalyst was air dried for two days. The dried photocatalyst was then subjected to XRD for phase analysis. The XRD analysis was carried out in a Rigaku diffractometer (Model D/MAX-B, Rigaku Co., Tokyo, Japan) equipped with Ni-filtered Cu Kα radiation (λ = 0.15409 nm) at 40kV and 30mA settings. The 2θ step size was 0.02° and the scan rate was 1.8° min^{-1}.

XPS Analysis

The air-dried photocatalyst after the photocatalytic degradation study was subjected to XPS analysis in a XPS-ESCA PHI 5400 spectrometer using Mg Kα radiation. The XPS spectra of Ti 2p and O 1s in TiO$_2$ were then recorded.

RESULTS AND DISSCUSSIONS

Photocatalytic activity

The photcatalytic oxidation of organic pollutants in aqueous suspensions follows the Hinshelwood model. This kind of reaction can be represented as [22, 23]:

$$-\frac{dC}{dt} = k_r \frac{KaC}{1+KaC} \qquad (2)$$

Where (-dC/dt) is the degradation rate of methylene blue, C is the methylene blue concentration in the solution, t is the reaction time, k_r is the reaction rate constant and K_a is the adsorption coefficient of the reactant. When C is very small, the product K_aC is negligible with respect to unity. Now, setting the Eq. (2) at the initial conditions of photocatalytic procedure, t=0, the concentration transforms to C = C$_o$, which gives the following equation:

$$-\text{Ln}\left(\frac{C}{C_o}\right) = kt \qquad (3)$$

Where k is the reaction rate constant. Figure 3 shows a linear correlation between Ln (C_o/C) and t for methylene photocatalytic degradation over TiO$_2$ and CuO-TiO$_2$ and the corresponding kinetics constant are shown in Table 1.

The photcatalytic activity is due to photo-induced electrons and the corresponding positive holes that are formed. The photocatalytic activity of nanocrystalline TiO$_2$ in this case is low, because we believe that the generated photoinduced holes easily recombine with photoinduced electron due to the narrow band energy gap, thus decreasing the photocatalytic efficiency. If the recombination rate is very fast, there will not be enough time for any reaction to occur which results in little or no photocatalytic effect. Hence, through CuO loading, we believe that the band gap energy of TiO$_2$ had been lowered, thus improving the photocatalysis activity. As seen in Figure 2, the maximum photocatalytic degradation activity was recorded for 20 wt% CuO-TiO$_2$. The photocatalytic activity increased with an increase in the amount of CuO up to 20 wt% loading, but further loading led to a decrease in the photocatalytic activity. The high photocatalytic activity of CuO-TiO$_2$ is due to the formation of Ti—O—Cu bond formed on the surface of TiO$_2$. The Cu^{2+} ions act as electron acceptors thus trapping the electrons in the conduction band of TiO$_2$. In this process, the electrons accumulate on the photocatalyst and the holes in the valence band oxidize OH$^-$, H$_2$O or the organic medium, accordingly avoiding the electron hole recombination. If the rate of electron-hole recombination is low, the photocatalytic activity will be higher and so more electrons trapped in the photocatalyst are transferred to adsorbed O$_2$. But as the percentage of CuO increases the photocatalytic activity decreases, as the excess of CuO depositing on TiO$_2$ will screen the photocatalyst from the UV source. The photocatalytic activity of 30 wt% CuO loaded TiO$_2$ is even lesser than pure TiO$_2$ due to the screening effect. Hence, 20 wt% is the optimum amount of CuO needed for good photocatalytic activity.

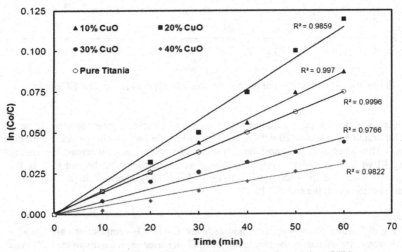

Figure 2. Effect of CuO on the photocatalytic activity of TiO$_2$

Table 1: Kinetics constant and half-life periods for CuO-loaded TiO$_2$

CuO deposited content (wt %.)	Kinetics constant (min^{-1})	Half-life - $t_{1/2}$ (h)
0	0.001282	9.0
10	0.001418	8.1
20	0.001752	6.6
30	0.000826	14.0
40	0.000442	26.1

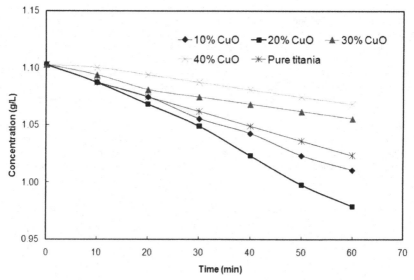

Figure 3. Photocatalytic degradation of methylene blue by copper oxide loaded TiO$_2$ under the action of UV light

Figure 3 shows the photodecomposition kinetics of methylene blue solution. It can be seen that the solution containing 20 wt% of CuO underwent the highest change in concentration with time. The concentration almost linearly decreased with time. Whereas, the solution containing 40 wt% CuO showed very little change in concentration at the end of 1 h, thus indicating the minimum photocatalytic activity. Pure TiO$_2$ showed linear decrease in concentration with time at the end of 1 h.

Phase Analysis
X-ray diffraction patterns of TiO$_2$ and different CuO-TiO$_2$ compositions are shown in Figure 4. It was observed that as the percentage of CuO increased the intensity of the CuO peaks also increased. The diffraction pattern did not indicate the formation of any secondary phases.

The presence of anatase and CuO phases were confirmed by comparing with JCPDS standard files # 21-1272 and # 80-1916, respectively. The crystallite size was calculated by the Scherrer equation:

$$\beta = \frac{0.9\lambda}{d\cos\theta}$$

(4)

Where λ is the wavelength of X-rays, θ is the braggs angle, d is the average crystallite size, β is the full width at half maximum. Using equation (4) the crystallite size of TiO$_2$ powder was found out to be 10.7 nm.

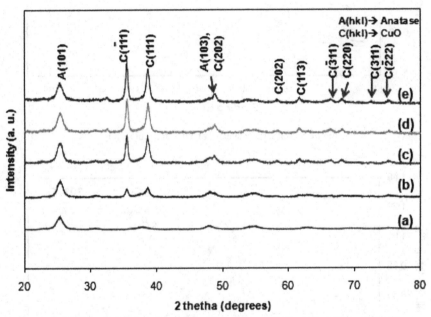

Figure 4. XRD analysis of photocatalysts after photocatalytic study (a) Pure TiO$_2$ (b) 10wt% CuO-TiO$_2$ (c) 20wt% CuO-TiO$_2$ (d) 30wt% CuO-TiO$_2$ (e) 40wt% CuO- TiO$_2$

XPS Analysis
The XPS Ti 2p spectra of TiO$_2$ and CuO-TiO$_2$ are as shown in Figure 5. The spectra show double peaks (Ti 2p$_{1/2}$ and Ti 2p$_{3/2}$). It can be seen that as the percentage of CuO increases the binding energy also increases. This shift in binding energy is due to the dispersion of copper on TiO$_2$ and the formation of Ti—O—Cu bond. The XPS O1s spectra for pure as well as CuO loaded TiO$_2$ is as shown in Figure 6. The O1s spectra also showed increasing binding energy values as the percentage of CuO increased. Thus, the increase in the binding energy of both Ti 2p and O 1s indicates the formation of Ti—O—Cu bond.

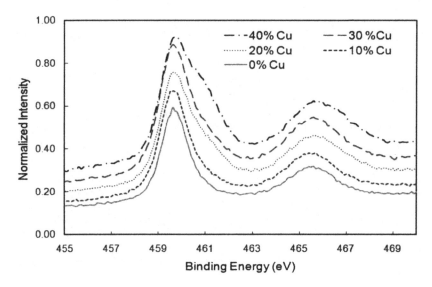

Figure 5. XPS Ti 2p spectra of TiO₂ and CuO-TiO₂

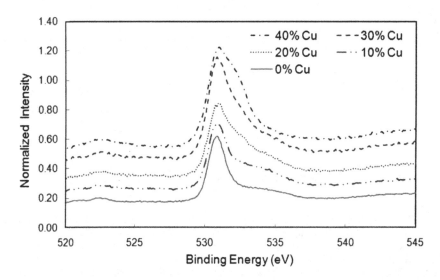

Figure 6. XPS O1s spectra of TiO₂ and CuO-TiO₂

Table 2. Elements and Spectral lines showing the binding energy and oxidation states (adapted from NIST X-ray Photoelectron Spectroscopy (XPS) Database)

Element	Formula	Oxidation state	Spectral line	Binding Energy (eV)
Ti	Ti	Zero	$2p_{1/2}$	460.0
			$2p_{3/2}$	453.2 – 454.3
	Ti$_2$O	+1	------	------
	TiO	+2	$2p_{1/2}$	460.2
			$2p_{3/2}$	455.1 – 455.9
	Ti$_2$O$_3$	+3	$2p_{1/2}$	462.0
			$2p_{3/2}$	456.8 – 457.8
	TiO$_2$	+4	$2p_{1/2}$	464.3 – 464.7
			$2p_{3/2}$	458.5 – 459.3
O	O$_2$	-2	1s	529.9 – 531.2
Cu	Cu	Zero	$2p_{1/2}$	932.2-932.66
			$2p_{3/2}$	932.6- 933.1
	Cu$_2$O	+1	$2p_{1/2}$	952.7
			$2p_{3/2}$	932.0-932.7
	CuO	+2	$2p_{1/2}$	952.7-953.7
			$2p_{3/2}$	932.9-934.1

The binding energy values of $2p_{1/2}$ and $2p_{3/2}$ of Ti in TiO$_2$ were compared to the values available in literature, which is as tabulated in Table 2. The binding energy values obtained in our case was close enough to the values in Table 2. In all the peaks shown in Figure 5 and 6, there was an overall shift of 1 eV, which is attributed to the carbon correction. The Figure 5 shows that the spin-orbit components (namely $2p_{3/2}$ and $2p_{1/2}$) is deconvoluted by two curves (approximately 459.5 eV and 465.5 eV), thus indicating that Ti mainly existed as Ti^{4+}. These values agreed well with the values available in literature [14, 24]. It is also observed that as the amount of CuO increases the peaks lose their symmetry and shifts towards higher binding energy values. However, the reason for this change has yet not been investigated.

XPS Cu 2p spectrum was not recorded, as Cu (0), Cu (I) and Cu (II) spectral lines cannot be distinguished from one another because their binding energy values are too close for accurate determination from XPS spectrum. This is clearly evident from the Table 2 as well. Hence, we did not study the oxidation behavior of Cu after photocatalysis.

CONCLUSIONS

The photocatalytic degradation of methylene blue by pure and CuO loaded nanocrystalline anatase was investigated. CuO loaded nanocrystalline anatase proved to be an efficient photocatalyst. The results demonstrated that 20 wt% CuO was the optimum quantity for maximum photocatalytic activity. CuO on the surface of TiO$_2$ can trap electrons from the TiO$_2$ conduction band, and the electrons trapped on the CuO–TiO$_2$ site are subsequently transferred to the surrounding adsorbed oxygen, thereby avoiding electron–hole recombination, and enhancing the photocatalytic activity. However, excessive quantity of copper (beyond 20 wt% CuO) screens off the photocatalyst from the UV radiation, resulting in noticeable decrease in activity. XRD analysis showed that the intensity of CuO peaks increased as the quantity of CuO in the photocatalyst increased. The average crystallite size was determined to be 10.7 nm using the Scherrer formula. XPS study showed that the binding energy of Ti 2p and O1s bands in different CuO-TiO$_2$ samples increased due to the atomic dispersion of Cu on TiO$_2$ and the formation of Ti—O—Cu bond. This study also concludes that the oxidation state of Ti in TiO$_2$ did not change at the end of photocatalysis.

REFERENCES
1.   Zhang, L., Kanki, T., Sano, N., and Toyoda, A., "Development of TiO$_2$ photocatalyst reaction for water purification". Separation and Purification Technology, 2003. 31(1): p. 105-110.
2.   Hagfeldt, A., Gratzel, M., "Light-induced redox reactions in nanocrystalline systems". Chem. Rev, 1995. 95: p. 49.
3.   Yeh, Y.C., Tseng, T.T., Chang, D.A., "Electrical properties of porous titania ceramic humidity sensors". J. Am. Ceram. Soc, 1989. 72: p. 1472.
4.   Balfour, J.G., "Technological Applications of Dispersions". Marcel Dekker: New York, 1994.
5.   Diebold, U., "The surface science of titanium dioxide". Surface Science Reports, 2003. 48(5-8): p. 53-229.
6.   Rao, M.V., Rajeshwar, K., Verneker, P., DuBow, J., "Photosynthetic production of hydrogen and hydrogen peroxide on semiconducting oxide grains in aqueous solutions". J. Phys. Chem, 1980. 84: p. 1987.
7.   Yuan, Z.H., Jia, J.H., Zhang, L.D., "Influence of co-doping of Zn(II)+Fe(III) on the photocatalytic activity of TiO$_2$ for phenol degradation". Materials Chemistry and Physics, 2002. 73(2-3): p. 323-326.
8.   Jakob, M., Levanon, H., Kamat, P.V., "Charge distribution between UV-irradiated TiO$_2$ and gold nanoparticles: Determination of shift in the fermi level". Nano Lett, 2003. 3: p. 353.
9.   Bae, E., Choi. W., "Highly enhanced photoreductive degradation of perchlorinated compounds on dye-sensitized metal/TiO$_2$ visible light". Environ. Sci. Technol., 2003. 37: p. 147.
10.   Uhm, Y.R., Woo, S.H., Kim, W.W., Kim, S.J., Rhee, C.K., "The characterization of magnetic and photo-catalytic properties of nanocrystalline Ni-doped TiO$_2$ powder synthesized by mechanical alloying". Journal of Magnetism and Magnetic Materials, 2006. 304(2): p. e781-e783.

11. Xu, Y.H., Liang, D.H., Liu, M.L., Liu, D.Z., "Preparation and characterization of Cu$_2$O-TiO$_2$: Efficient photocatalytic degradation of methylene blue". Materials Research Bulletin, 2008. 43(12): p. 3474-3482.

12. Liu, G., Zhang, X., Xu, Y., Niu, X., Zheng, L., Ding, X., "The preparation of Zn^{2+}-doped TiO$_2$ nanoparticles by sol-gel and solid phase reaction methods respectively and their photocatalytic activities". Chemosphere, 2005. 59(9): p. 1367-1371.

13. Fan, X., Chen, X., Zhu, S., Li, Z., Yu, T., Ye, J., Zou, Z., "The structural, physical and photocatalytic properties of the mesoporous Cr-doped TiO$_2$". Journal of Molecular Catalysis A: Chemical, 2008. 284(1-2): p. 155-160.

14. Xu, Y.H., Chen, H.R., Zeng, Z.X., Lei, B., "Investigation on mechanism of photocatalytic activity enhancement of nanometer cerium-doped titania". Applied Surface Science, 2006. 252(24): p. 8565-8570.

15. Yu, J., Liu, S., Xiu, Z., Yu, W., Feng, G. "Combustion synthesis and photocatalytic activities of Bi^{3+} doped TiO$_2$ nanocrystals". Journal of Alloys and Compounds, 2008. 461(1-2): p. L17-L19.

16. Shi, J.W., Zheng, J.T., Hu, Y., Zhao, Y. "Influence of Fe^{3+} and Ho^{3+} co-doping on the photocatalytic activity of TiO$_2$". Materials Chemistry and Physics, 2007. 106(2-3): p. 247-249.

17. Wang, H.W., Lin, H.C., Kuo, C.H., Cheng, Y.L., Yeh, Y.C., "Synthesis and photocatalysis of mesoporous anatase TiO$_2$ powders incorporated Ag nanoparticles". Journal of Physics and Chemistry of Solids, 2008. 69(2-3): p. 633-636.

18. Xin, B., Jing, L., Ren, Z., Wang, B., Fu, H., "Effects of simultaneously doped and deposited Ag on the photocatalytic activity and surface states of TiO$_2$". J. Phys. Chem. B, 2005. 109: p. 2805.

19. Qiu, S., Kalita, S.J., "Synthesis, processing and characterization of nanocrystalline titanium dioxide". Materials Science and Engineering: A, 2006. 435-436: p. 327-332.

20. Foster, N.S., Noble, R.D., Koval, C.A., "Reversible photoreductive deposition and oxidative dissolution of copper ions in titanium dioxide aqueous suspensions". Environ. Sci. Technol., 1993. 27: p. 350.

21. Reiche, H.D.W., Bard, A.J., "Heterogeneous photocatalytic and photosynthetic deposition of copper on TiO$_2$ and WO$_3$ powders". J. Phys. Chem, 1979. 83: p. 2248.

22. Hong, S.S., Lee, M.S., Lee, G.D., Lim, K.T., Ha, B.J., "Synthesis of titanium dioxides in water-in-carbon dioxide microemulsion and their photocatalytic activity". Materials Letters, 2003. 57(19): p. 2975-2979.

23. Sakkas, V.A., Arabatzis, I.M., Konstantinou, I.K., Dimou, A.D., "Metolachlor photocatalytic degradation using TiO$_2$ photocatalysts". Applied Catalysis B: Environmental, 2004. 49(3): p. 195-205.

24. Losito, I., Amorisco A., Palmisano, F., Zambonin, P.G., "X-ray photoelectron spectroscopy characterization of composite TiO$_2$-poly(vinylidenefluoride) films synthesised for applications in pesticide photocatalytic degradation". Applied Surface Science, 2005. 240(1-4): p. 180-188.

CONSTITUENT PHASES OF NANOSIZED ALUMINA POWDERS SYNTHESIZED BY PULSED WIRE DISCHARGE

Satoru Ishihara [1], Yoshinori Tokoi [2], Yuu. Shikoda [1], Hisayuki Suematsu [1], Tsuneo Suzuki [1], Tadachika Nakayama [1], and Koichi Niihara [1]

[1] Extreme Energy-Density Research Institute, Nagaoka University of Technology, 1603-1 Kamitomioka, Nagaoka, Niigata 940-2188, Japan
[2] Department of Electrical and Electronic Engineering, Niigata University, 2-8050 Ikarashi Nishiku, Niigata, 950-2181, Japan

ABSTRACT

Pulsed wire discharge (PWD) method was developed to synthesize nanosized particles. In this method, pulsed large electric current goes through a thin metal wire, and then the wire can be instantaneously evaporated. The evaporated metal is quickly cooled in the ambient gas to form nanosized particles. In case that oxygen or air is used as the ambient gas, the evaporated metal is oxidized and results in formation of oxide particles. In this paper, results of Al_2O_3 powders by PWD experiments of Al wires at various conditions are demonstrated. The effects of experimental conditions on constituent phases in the synthesized powders were discussed. While the experiments to investigate the effects of charging energy and oxygen pressure were carried out in closed oxygen atmosphere, the effect of air flow was examined under regulated evacuation of air. The capacitors were charged by using a high-voltage DC power supply. Pulsed high current was applied by closing the spark gap. The formed particles in the chamber were evacuated and collected on a membrane filter. The obtained powders synthesized at most of the experimental conditions were composed of γ-Al_2O_3 and δ-Al_2O_3 phases. Instead of these phases, XRD peaks by another phase which was considered as θ-Al_2O_3 were detected in the powders synthesized at the lower pressures of 5 and 10 kPa. The γ/δ ratios of detected Al_2O_3 phases increased with increasing oxygen pressure and charged electric energy. The γ/δ ratios of Al_2O_3 phases synthesized in air flow conditions decreased with increasing air flow rate.

INTRODUCTION

Nanostructured materials and nanocomposites are highlighted in fields of materials science and technology, because such materials with structural scales in the range of 1-100 nm offer outstanding mechanical and multifunctional properties [1,2]. Powders with the particle sizes in such scales are referred to as nanosized powders. Size effect of decrease in the particle diameter to nanometer scales and the consequent increasing in the specific surface area bring some unique functions and properties [3]. Since Al_2O_3 ceramics are most widely used for industrial applications, nanosized Al_2O_3 powders have been much investigated from the standpoints of the sintering property and the catalytic activity. Besides nanosized powders, nanocomposite materials have also been reported to be consolidated by sintering from mixtures of nanosized powders. As an example, complex-shaped Al_2O_3/Ni nanocomposites with preserved multifunctional magnetic properties were successfully fabricated by a gelcasting process [4]. Thus, it is important to establish efficient processes to synthesize nanosized Al_2O_3 powders.

Various kinds of processes have been applied to synthesize nanosized powders. These processes can be classified into liquid phase methods and gas phase methods according to the principle of particle formation. Exploding wire (EW) and pulsed wire discharge (PWD) methods which are categorized in the gas phase methods were developed to synthesize nanosized powders. In these methods, large electric current goes through a thin metal wire, and then the wire can be instantaneously evaporated by the Joule's heat caused by the current. High efficiency of the direct energy conversion from electricity to heat is one of the principal advantages against the other gas phase methods. The evaporated metal is quickly cooled in the ambient gas to form nanosized particles. The history of EW and PWD methods date away back to 1857, in which M. Faraday reported that Ag, Cu, and Al powders were prepared by heating of metal wires by electric current [5]. After a long interval, this method was rediscovered and named as exploding wire method in 1950s [6]. Since then, research on EW method became active and was applied for preparing nanosized particles [7, 8]. Whereas nanosized metal particles are formed by evaporation of metallic wires in vacuum or inert gas atmosphere, the evaporated metal is oxidized and results in formation of nanosized oxide particles in case that oxygen or air is used as the ambient gas. Nanosized Al_2O_3 powders have been synthesized by EW and PWD methods of Al wires in oxygen and air atmospheres [9-13].

The Al_2O_3 system is known to involve a number of metastable polymorphic forms such as γ, δ, θ, η, κ, χ, besides the thermodynamically stable phase α (corundum) [14]. Al_2O_3 particles synthesized by most of gas phase methods have been reported to consist predominantly of γ-Al_2O_3 phase and partly of other metastable phases such as δ-Al_2O_3 and θ-Al_2O_3, rather than the most stable α-Al_2O_3 form [15, 16]. In previous studies for synthesizing nanosized TiO_2 powders by PWD, obtained powders included rutile and anatase phases, and the rutile phase content decreased with increasing relative energy which is defined as the ratio of the charged energy in the capacitor to the vaporization energy of the wire [17]. Such control of constituent phases in Al_2O_3 powders is important in view of suitable synthesizing for individual industrial application, since the polymorphic Al_2O_3 forms have much different characteristics. Thus, effects of synthesizing conditions on the constituent phases should be investigated by systematical experiments. In this paper, results of Al_2O_3 powders synthesized by PWD at various conditions are demonstrated. The effects of these experimental conditions on constituent phases in the synthesized powders were discussed.

EXPERIMENTAL DETAILS

A schematic diagram of the PWD apparatus used in this study is shown in Fig.1. Powder synthesis experiments were carried out according to the following procedure. A piece of pure Al wire was set between the electrodes (the nominal span length: 25 mm) in the chamber for each batch of experiment. The chamber was evacuated by the rotary pump and filled with oxygen gas to the prescribed pressures, in order to investigate the effects of oxygen pressure and charged energy. The effect of air flow was examined under regulated evacuation of air, by which the gas inlet was released and the gate valve was full- or partially- opened. The capacitor bank was charged by using a high-voltage DC power supply. Pulsed high current was applied by closing the spark gap. During each discharge experiment, the voltage between the electrodes and the current of the discharge circuit were measured with a digitized oscilloscope through two high-voltage probes and a current transformer. After each wire explosion, the formed particles in the chamber were evacuated and collected on a membrane filter with 0.1 μm pore size. The collected powders were analyzed by X-ray diffraction (XRD; Rigaku, RINT2000, Cu-Kα, 50 kV, 300 mA).

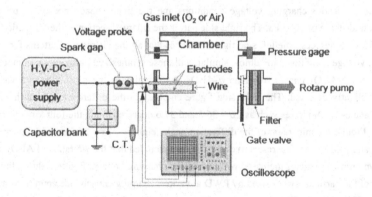

Figure 1. Schematic diagram of the pulsed wire discharge (PWD) apparatus.

The experimental conditions adopted in this study are summarized in Table I. The vaporization energy (E_v) of pure Al wire was calculated from the reference data of the enthalpy change up to vaporization at the boiling point [18], and the volume of wire (diameter: 0.2 mm and 0.1mm, nominal length: 25 mm). The charged energy, E_c can be calculated by the equation;

$$E_c = (1/2) \; C \; V_c^2 \tag{1},$$

where, C is the capacitance of the capacitor bank and V_c is the charging voltage, respectively. Consequently, the relative energy ($K = E_c/E_v$) in this study varied between $K = 1.5$ and 48.

Table I. Experimental conditions

Wire	Al, ϕ 0.2, 25 mm						Al, ϕ 0.1, 25 mm
Vaporization energy, E_v	29.9 J						7.5 J
Atmosphere pressure, p	(5 kPa), 10 kPa, 50 kPa, 100 kPa						10 kPa, 100 kPa
Capacitance, C	10 μF				20 μF	30 μF	20 μF
Charging voltage, V_c	3 kV	4 kV	5 kV	6 kV	6 kV	6 kV	6 kV
Charged energy, E_c	45 J	80 J	125 J	180 J	360 J	540 J	360 J
Relative energy, $K = E_c/E_v$	1.5	2.7	4.2	6.0	12	18	48

RESULTS AND DISCUSSION

(1) Effects of charged energy and oxygen pressure

PWD experiments of Al wires were carried out in closed oxygen atmospheres, in order to investigate the effects of charged energy and oxygen pressure. Figure 2 shows the XRD patterns of the particles synthesized by the PWD experiments of pure Al wires in diameter of ϕ 0.2 mm at various charging voltage conditions and the fixed capacitance of 10 μF in the oxygen atmosphere of 100 kPa. The right side part is the enlarged section at the diffraction angle range of 30~45 degrees. The left side figure indicates that all the diffraction patterns for various charging voltage conditions are almost similar, and these synthesized particles are composed of γ-Al$_2$O$_3$ and δ-Al$_2$O$_3$ phases, whereas α-Al$_2$O$_3$ phase which is the most thermodynamically stable is slightly detected. The right side figure emphasizes the difference in peak heights of γ-Al$_2$O$_3$ and δ-Al$_2$O$_3$ phases such as $2\theta = 36.5$ and 37.6 degrees among the four kind diffraction patterns. From the comparison of the diffraction peak heights of $2\theta = 36.5$ and 37.6 degrees for δ-Al$_2$O$_3$ and γ-Al$_2$O$_3$ phases, respectively, it can be concluded that the γ/δ ratios of Al$_2$O$_3$ phases in the synthesized particles increased with increasing charging voltage. Figure 3 shows the XRD patterns of the particles synthesized by PWD at various charging capacitance conditions and the fixed charging voltage of 6 kV. This result indicates that the γ/δ ratios of Al$_2$O$_3$ phases increased with increasing charging capacitance.

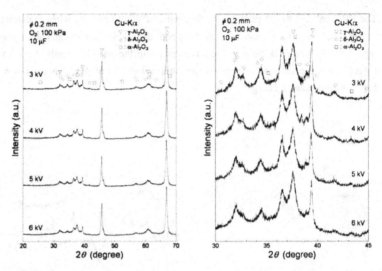

Figure 2. XRD patterns of the particles synthesized by PWD at various charging voltage conditions in the oxygen atmosphere of 100 kPa. The right side part is enlarged at the range of 30~45 degrees.

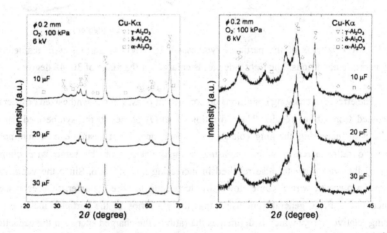

Figure 3. XRD patterns of the particles synthesized by PWD at various capacitance conditions in the oxygen atmosphere of 100 kPa. The right side part is enlarged at the range of 30~45 degrees.

The XRD patterns of the particles synthesized by PWD of pure Al wires in diameters of $\phi 0.1$ mm and $\phi 0.2$ mm are compared in Fig. 4. In both the cases of the oxygen atmospheres of 10 kPa and 100 kPa, the peak height of δ-Al_2O_3 phase in the particles synthesized from the wires of $\phi 0.1$ mm was much less detected in comparison with that of $\phi 0.2$ mm as shown in the right side part for the enlarged range.

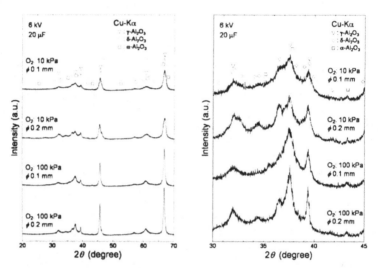

Figure 4. XRD patterns of the particles synthesized by PWD at various wire diameter and oxygen pressure conditions. The right side part is enlarged at the range of 30~45 degrees.

The effects of charging conditions on constituent phases in the synthesized powders can be discussed from above results. The γ/δ ratios of Al_2O_3 phases in the synthesized particles increased with increasing charging voltage and capacitance. These results can be summarized that the γ/δ ratio increases with increasing charged energy for the same wire diameter. Decreasing the wire diameter also resulted in increasing the γ/δ ratio. Since the vaporization energy per a piece of wire becomes lower by decreasing in wire diameter, this result can be combined into a more general conclusion that the γ/δ ratios in the particles increases with increasing relative energy which is defined as the ratio of the charged energy in the capacitor to the vaporization energy of the wire.

In the previously reported results of synthesizing nanosized TiO_2 powders by PWD of pure Ti wires in oxygen atmosphere, rutile and anatase phases were formed, and the rutile phase content decreased with increasing relative energy [17]. This result is simply considered that the higher temperature of vapor can be obtained by the higher charged energy, since rutile phase is more stable at higher temperatures than anatase phase. However, the result of Al_2O_3 in this study does not agree with that of TiO_2, since δ-Al_2O_3 phase is usually more stable at higher temperatures than γ-Al_2O_3 phase. This reason is still unclear. It may be explained by the relative thermal stability change in nanosized powders, which is caused by effects of increased surface area [19].

Figure 5 shows the XRD patterns of the particles synthesized by PWD at the charging conditions of 3 kV and 10 μF in oxygen atmosphere of various pressures. At the lowest pressure of 5 kPa, diffraction peaks of γ-Al_2O_3 and δ-Al_2O_3 phases are less detected in comparison with that at 100 kPa as described above. Instead of these phases, some different diffraction peaks by another phase are detected. It is considered as θ-Al_2O_3 phase, although these diffraction angles imperfectly match the standard data of θ-Al_2O_3. This phase is less detected and δ-Al_2O_3 phase becomes dominant at the intermediate pressure of 50 kPa. The γ/δ ratios of Al_2O_3 phases increased with increasing oxygen pressure at higher pressures.

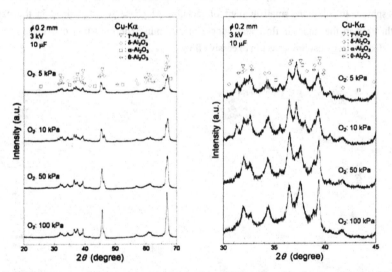

Figure 5. XRD patterns of the particles synthesized by PWD at the charging conditions of 3 kV and 10 μF in oxygen atmosphere of various pressures. The right side part is enlarged at the range of 30~45 degrees.

The effect of oxygen pressure on the synthesis of Al_2O_3 powders were investigated in our previous study (Ref. 11). However, δ-phase and θ-phase were not identified in the previous study which was measured by XRD at an X-ray condition of 40 kV-30 mA, while the present XRD study used a stronger X-ray at 50 kV-300 mA. Although TEM is often used for crystalline investigation, typical difference in TEM selected area electron diffraction analyses was not found among the Al_2O_3 powders synthesized at different pressure conditions. In this study, high-power XRD is more effective to investigate the constituent phases than TEM, since the crystalline structures of Al_2O_3 phases are relatively complicated.

(2) Effect of air flow atmosphere

Air is more suitable atmosphere than oxygen for production of nanosized Al_2O_3 powders by PWD in industrial applications taking into account of the cost and safety. Then, PWD experiments were tried in air flow atmospheres. The air flow rates were evaluated by a gas flow meter as 33L/min and 15L/min for the conditions of the full- and partially- opened gate valve, respectively. Figure 6 shows XRD patterns of the particles synthesized by PWD in various conditions of air flow atmosphere. The obtained particles in all experimental conditions were mostly composed of γ-Al_2O_3 and δ-Al_2O_3 phases as same as the results in the high oxygen atmosphere. In addition, small amount of metallic Al phase was remained in the particles synthesized in the high air flow atmosphere of 33L/min. The oxidation of Al vapor may be imperfect by large cooling effect in excessive flow rate of air.

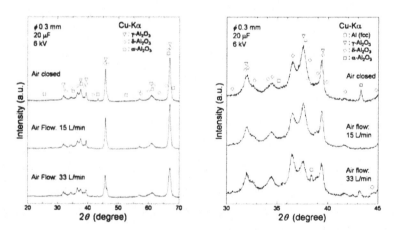

Figure 6. XRD patterns of the particles synthesized by PWD in various conditions of air flow atmosphere. The right side is enlarged at the range of 30~45 degrees.

The γ/δ ratios of the Al_2O_3 phases in the obtained particles decreased with increasing air flow rate at the same charging conditions. This tendency is same as that in the case of decreasing oxygen pressure conditions. The increase of air flow is considered to affect the PWD phenomena as dilution of evaporated phase of Al as similar to lower pressure conditions in the oxygen atmosphere conditions in which the metal vapor can expand easily by the pressure difference between the vapor and the atmosphere.

CONCLUSIONS

The synthesized particles at most of experimental conditions were mainly composed of γ-Al_2O_3 and δ-Al_2O_3 phases. The γ/δ ratios of Al_2O_3 phases in these particles increased with increasing charged energy and oxygen pressure. Instead of these phases, another phase which was considered as θ-Al_2O_3 was detected at the conditions of low pressures of 5 and 10 kPa. The γ/δ ratios of Al_2O_3 phases formed in air flow conditions decreased with increasing air flow rate, as similar to lower pressure conditions in oxygen atmosphere.

ACKNOWLEDGMENT

This study was financially supported by Strategic Information and Communications R&D Promotion Programme (SCOPE), Ministry of Internal Affairs and Communications, Japan.

REFERENCES
[1] H. Gleiter, Nanostructured materials: state of the art and perspectives, *Nanostructured Materials*, **6**, 3-14 (1995).

[2] K. Niihara, New design concept of structural ceramics nanocomposites, *J. Ceram. Soc. Japan*, **99**, 974-982 (1991).

[3] F. E. Kruis, H. Fissan and A. Peled, Synthesis of nanoparticles in the gas phase for electronic, optical and magnetic applications - A review, *J. Aerosol Sci.*, **29**, 511-535, (1998).

[4] K. Niihara, B.-S. Kim, T. Nakayama, T. Kusunose, T. Nomoto, A. Hikasa, T. Sekino, Fabrication of complex-shaped alumina/nickel nanocomposites by gelcasting process, *J. Euro. Ceram. Soc.*, **24**, 3419-3425 (2004).

[5] M. Faraday, The bakerian lecture: Experimental relations of gold (and other metals) to light, *Phil. Trans. Royal Soc. London*, **147**, 145-181 (1857).

[6] *Exploding wires* (edited by W. G. Chace and H. K. More), **1**, (1959).

[7] F. G. Karioris and B. R. Fish, An exploding wire aerosol generator, *J. Colloid Sci.*, **17**, 155-161 (1962).

[8] W. Jiang and K. Yatsui, Pulsed wire discharge for nanosize powder synthesis, *IEEE Trans. Plasma Sci.*, **26**, 1498-1501 (1998).

[9] M. Umakoshi, T. Yoshitomi, and A. Kato, Preparation of alumina and alumina-silica powders by wire explosion resulting from electric discharge, *J. Mater. Sci.*, **30**, 1240-1244 (1995).

[10] V. Ivanov, Y. A. Kotov, O. H. Samatov, R. Böhme, H. U. Karow, and G. Schumacher, Synthesis and dynamic compaction of ceramic nano powders by techniques based on electric pulsed power, *Nano Struct. Mater.*, **6**, 287-290 (1995).

[11] T. Suzuki, K. Keawchai, W. Jiang and K. Yatsui, Nanosize Al_2O_3 Powder production by pulsed wire discharge, *Jpn. J. Appl. Phys.*, **40**, 1073-1075 (2001).

[12] V. Sabari Giri, R. Sarathi, S. R. Chakravarthy, and C. Venkataseshaiah, Studies on production and characterization of nano-Al_2O_3 powder using wire explosion technique, *Mater. Lett.*, **58**, 1047-1050 (2004).

[13] R. Baksht, A. Pokryvailo, Y. Yankelevich, and I. Ziv, Explosion of thin aluminum foils in air, *J. Appl. Phys.*, **96**, 6061-6065 (2004).

[14] I. Levin, and D. Brandon, Metastable alumina polymorphs: Crystal structures and transition sequences, *J. Am. Ceram. Soc.*, **81**, 1995-2012 (1998).

[15] L. Fu, L. D. Johnson, J. G. Zheng, and V. P. Dravid, Microwave plasma synthesis of nanostructured γ-Al_2O_3 powders, *J. Am. Ceram. Soc.*, **86**, 1635-1637 (2003).

[16] K. Suresh, V. Selvarajan, and M. Vijay, Synthesis of nanophase alumina, and spheroidization of alumina particles, and phase transition studies through DC thermal plasma processing, *Vacuum*, **82**, 814-820 (2008).

[17] Y. Tokoi, T. Suzuki, T. Nakayama, H. Suematsu, W. Jiang, and K. Niihara, Synthesis of TiO_2 nanosized powder by pulsed wire discharge, *Jpn. J. Appl. Phys.*, **47**, 760-763 (2008).

[18] Standard formation enthalpy of gas from NIST Chemistry Webbook, http://webbook.nist.gov/chemistry/

[19] J. M. McHale, A. Navrotsky, and A. J. Perrotta, Effects of increased surface area and chemisorbed H_2O on the relative stability of nanocrystalline γ-Al_2O_3 and α-Al_2O_3, *J. Phys. Chem.*, **B, 101**, 603-613 (1997).

THE FORMATION OF NANOSTRUCTURE COMPOUND LAYER DURING SULFUR PLASMA NITRIDING AND ITS MECHNICAL PROPERTIES

Kyoung Il Moon[1], Yoon Kee Kim[2], Kyung Sub Lee[3]

[1] Korea Institute of Industrial Technology, Technology Service Division for SMEs, Incheon technology Service Center, Incheon 406-840, Korea, kimoon@kitech.re.kr

[2] Hanbat National University, Dept. of Welding & Production Engineering, SAN16-1, DuckMyoung-dong, Yuseon-gu, Daejeon 305-719, Korea

[3] Hanyang University, Division of Materials Science and Engineering, Seoul 133-791, South Korea

ABSTRACTS

Plasma nitriding process is performed generally with N_2 and C_3H_8 gases to diffuse N and C atoms into steels. In this study, it has been investigated on the microstructure change during plasma nitriding with the addition of C and S source gases of H_2S and C_3H_8. It has been surveyed the effect of the microstructural change on the mechanical properties. The substrate materials were S45C. The process temperatures were 450-550 °C. The working pressures were 2 torr. The amount of H_2S gas was ranged from 5 to 35 SCCM and C_3H_8 was 100 to 500 SCCM. XRD, SEM, EDS analyses were performed to investigate the structural changes and micro-hardness and wear tests were examined to find out the mechanical properties.

With increasing H_2S gas, the thickness of compound layer was increased and the brittle and porous FeS layer with the grain size of 500 nm that was formed on the top of the compound layer. The hardness increased to some point and then decreased with increasing H_2S gas. The data on the friction coefficient showed the lowest value of 0.4 at the highest hardness when only H_2S gas added. The addition of C_3H_8 with H_2S gas made the FeS layer turned into a dense and hard structure but the compound layer had a large grain size over 300 nm of compound layer. However, it is found that process relevant factors could be controlled to restrict the growth of the compound layers, the compound layer could be maintained as nano-sized less than 50 nm. Increase of carbon content resulted in amorphous-like compound layer. Moreover the formation of such metastable structures and a fine distribution of FeS particles improved the wear property. The best friction coefficient was obtained in this specimen and less than 0.2.

INTRODUCTION

Automobile and machinery components require good tribological properties to ensure long life time and good efficiency. Especially, it is attractive that a low friction of automobile parts, especially motion parts, could be an effective method to improve fuel efficiency. Nitriding is one of the widely used surface hardening treatment technologies for automobile and machinery and it provides high hardness with good wear properties [1-5]. During the nitriding, nitrogen atoms are introduced into the surface of steels at a temperature range (500 to 550 °C, or 930 to 1020 °F). Thus, nitriding is similar to carburizing in that surface composition is altered, but different in that nitrogen is added into ferrite instead of austenite. Because nitriding does not involve heating into the austenite phase field and a subsequent quench to form martensite, nitriding can be accomplished with a minimum of distortion and with excellent dimensional control. Nitrogen has a very small amount of solubility in iron. It can form a solid solution with ferrite at nitrogen contents up to about 0.1 %. At above 0.1% N, a compound called gamma prime (γ'), with a composition of Fe_4N is formed. At nitrogen contents greater than 6%, the equilibrium reaction product is ε compound, $Fe_{2-3}N$. The ε zone of the case is hardened by the formation of the $Fe_{2-3}N$ compound, and below this layer there is some solid solution strengthening from the nitrogen in solid solution. However, the compound layer that is generally formed by conventional gas nitriding or plasma nitriding has a high friction coefficient, which may be harmful on the wear property of machine, especially on the counting parts. As an effective process to improve wear properties of mechanical components such as bushing and gear, plasma sulfur nitriding has great attractions. This process leads to the formation of a very thin FeS iron sulfide layer on the top of a thick compound layer. The FeS layer could be an excellent lubricant due to its hexagonal structure. So, the existence of FeS layer results in wear resistance and anti-seizure properties of mechanical parts [6]. According to the results on the sulfur nitriding processes, the best friction coefficient data was 0.3. But it has not been proposed the reason well for the low friction coefficient of such sulfur nitriding process. Also, the wear properties of materials should be dependent on the surface properties such as morphology and hardness, roughness. However, up to now, there have been few reports on the relation between the surface microstructure and the wear properties. The objective of the present study is to find out the best sulfur nitriding conditions by investigation the surface microstructure changes and its effects on the tribological properties.

Experimental details

Samples were S45C steel that is most widely used for automobile and machinery parts. S45C samples were cut from 30 mm diameter bar to disc of 10 mm thickness. The sample experienced quenching and tempering after normalizing. The sample surface was polished and cleaned in ultrasonic bath. Then samples were loaded in the plasma nitriding equipment shown in Fig. 1. The characteristic feature of this machine is that there is a subsidiary cathode surround the working zone. By this subsidiary plasma generator, nitriding is possible without plasma directly charging on the samples. Except many advantages such as high productivity and easy charges of work loads with different geometry and size, and avoiding the common problems associated with conventional ion nitriding [1,2], the chamber has many outstanding properties as follows;

1. Uniform distribution of temperature → high productivity
2. Extracted plasma from electron
3. Easy control of case depth, nitriding structure with & without compound layer
4. Rapid nitriding both with bias cathode & subsidiary cathode
5. High surface quality (surface roughness less than 0.02 μm)
6. Possibility of nitriding & DLC coating in a chamber (easy combination with PECVD process)

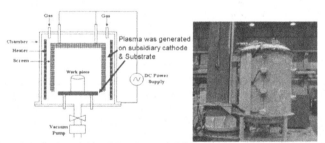

Figure 1. Schematic drawing of the plasma nitriding chamber and its real photograph

As mentioned in number 3, the nitriding is possible without compound layer by using only subsidiary cathode. In this case, very fine microstructure was formed on the surface and depth profile was measured as 120 μm after 2 hours nitriding. Also, by using both subsidiary cathode and bias cathode, compound layer and case depth could be formed very rapidly and they were over 20 μm and 0.5 μm, respectively within only 2 hours. Most of all, the chamber has enough size to be charged with the samples of 800 kg.

In this study, sulfur nitriding was performed with bias and subsidiary cathodes connected to the DC pulse power supply with the capacities of 50 kW & 20 kW, respectively. The base pressure for sulfur nitriding was 6×10^{-3} torr that was attained by rotary pump and booster pump, and temperature was 450~550 °C. Before main process, the sample was precleaned with Ar/H_2 plasma at the pressure of 1.25 torr for 1 hour during the heating process. The bias voltage of the pre-cleaning was – 600~ 700V. For main process of sulfur nitriding, N_2, H_2 and 5 % H_2S/N_2 mixed gases were introduced with the ratio of $H_2 : N_2 : 5\% H_2S/N_2 = 1:1:0.05 \sim 1:1:0.5$. The nitrogen gas and the hydrogen gas were fixed as 2000 SCCM, respectively. Thus, 5 % H_2S/N_2 mixed gas was added from 100 to 1000 SCCM. But the amount H_2S gas was only 5 to 50 SCCM because of their percentage was 5. The process pressure was 2 torr. In additions, 2~10 % C_3H_8 gas of total gas was added to examine hydrocarbon gas effect. Plasma sulfur nitriding was performed for 2 hours.

The surface hardness and case depth were measured with micro Vickers hardness tester (Future Tech FM-7). Microstructure of sulfur nitriding layer was observed by optical microscope (Z16APO) and scanning electron microscope (SEM, HITACH S-4300). Surface roughness was observed with atomic force microscope (AFM, NS4A). X-ray diffraction (XRD, Rigaku RAD-3C) was used for phase analysis of sulfur nitriding layer. Chemical composition was analyzed using electron probe micro analyzer (EPMA, Shimadzu EPMA-1400).

Friction coefficient was measured by ball on disc method. For ball on disc test, 6 mm-diameter SUJ2 ball was used. The test temperature was room temperature and humidity was about 40 ~50 %. The friction coefficient was measured up to 1000 m. Rotating speed was 5 cm/s and test load was 5 N. After the test, wear amount was calculated from wear track width and the wear track was observed with SEM.

RESULTS AND DISCUSSION

Figure 3 are optical micrographs and SEM micrographs that shows the surface morphology changes of the plasma sulfur nitriding samples with increasing H_2S/N_2 mixed gas. According to the optical micrographs, black area was increase with increased H_2S/N_2 mixed gas and it was fully covered the surface with 1000 sccm mixed gas. SEM micrographs showed that the black area was porous structure as indicated in the black circle of the far right SEM picture. This porous structure was increase with the mixed gas ratio and it was fully covered the surface with 1000 sccm mixed gas. The formation of such porous layer was also reported in several researches [6-9]. They reported that the formation of such porous layer resulted in the decrease of the surface hardness. Thus, in this study, it is expected that such porous layer resulted in the decrease of the surface hardness.

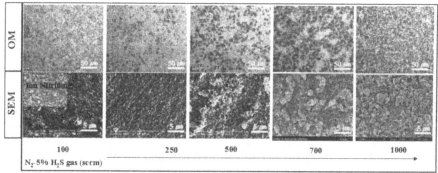

Figure 2. Optical micrographs and SEM micrographs of plasma sulfur nitriding S45C with increasing H_2S/N_2 mixed gas.

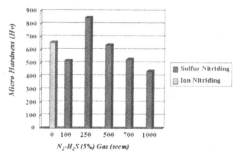

Figure 3. Surface hardness changes with H_2S/N_2 mixed gas.

Figure 3 shows the changes of surface micro-hardness with H_2S/N_2 mixed gas. According to the micro-hardness test, the surface hardness was increased up to 850 Hv with 250 sccm mixed gas and then, they decreased with further addition of H_2S/N_2 mixed gas. Especially, the surface microstructure that was fully covered with porous structure had not good hardness less than 500 Hv. Even, the surface layer was easily removed by small shear stress. According to surface hardness data, the samples prepared with 100 sccm and 700 sccm mixed gas addition had almost same hardness although they had different surface structure with the different portion of porous area. So, in this study, it has been studied in detail on these two samples. For this, 4 kinds of samples were prepared with plasma nitro-carburizing for 4 hours at 500 °C. The sample 1 and 2 were prepared with the gas compositions of 100 and 700 sccm mixed gas that were decided from the surface hardness data. Actually, compound layer could be formed easily by addition of C-source gas such as propane and methane. Thus, 100 sccm C_3H_8 gas was added for samples 3 and 4 to investigate the effect of hydrocarbon gas. The specific conditions are shown in Table I.

Table I. The properties of the plasma sulfur nitriding S45C specimens

Samples	Conditions			C. D. 1) [μm]	C. L. 2) [μm]	Hv	Ra [μm]
	Temp	Time	Gas				
Ion nitriding (I. N.)	500 °C	4 h	50%N₂ + 50%H₂	225	7-15	680	0.15
1	500 °C	4 h	I.N.+0.1 % H₂S	225	9-15	730	0.07
2	500 °C	4 h	I.N.+0.7 % H₂S	225	20-25	400	0.13
3	500 °C	4 h	No.1+0.1% C₃H₈	225	4-5	900	0.07
4	500 °C	4 h	No.1+0.1% C₃H₈	225	4-5	800	0.13

1) C.D.: Case depth of the sample, 2) C.L.: compound layer

The thickness of nitriding layer was observed with optical microscope as shown in figure 4. The compound layer thickness of sample 1 was about 15 μm and that of sample 2 was 25 μm. Thus, with the increase H_2S gas, the thickness of the compound layer was increased. When C_3H_8 gas was introduced as in sample 3 and sample 4, only 4~5μm compound layer was formed on the surface. The results indicated that C_3H_8 gas addition during sulfur nitriding lower the rate of compound layer formation. Generally the C atoms activated the formation of compound layer during nitriding. The thick compound layer at the early stage of nitriding process must be resulted in the thin compound layer in the sample with C-source samples.

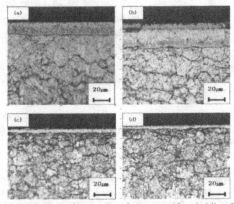

Figure 4. Cross-sectional optical micrographs of plasma sulfur nitriding S45C;
(a) sample 1, (b) sample 2, (c) sample 3, and (d) sample 4

Interestingly, double layers could be observed in compound layer of the sulfur nitriding samples and this double layer was most definite in the sample 2. In the sample 1, because the porous structure was only 20 % of the surface as shown in the second OM of figure 2, it was not easy to observe the double layer as in the sample 2. The case depth was determined by measurement of Vickers hardness profile from the surface. The case depth was determined up to the hardness was 50 Hv harder than the basal hardness. According to the data, all the samples had almost same case depth as 225 μm irrelevant of the amount of H_2S gas & C_3H_8 gas addition. This is because that the case depth was dependent on the diffusion time. The increase of the sulfur nitriding time resulted in the change of the surface microstructure and this resulted in the changes of surface hardness of the samples 1 and 2, according to comparison of the data in figure 4 and table 2. For samples 1 and 2, the increase of the sulfur nitriding time resulted in the increased of the compound layer thickness. This increased the surface hardness from 500 Hv to 730 Hv for the sample 1 with 100 sccm mixed gas. But for the sampe 2 prepared with 700 sccm mixed gas, the increase of the sulfur nitriding time increased both the compound layer and the porous layer thickness. This decreased the surface hardness from 500 Hv to 400 Hv. The other researcher [9] also reported that the porous layer after sulfur nitriding had low hardness of about 300 Hv.

Surface roughness of plasma sulfur nitriding S45C samples was measured by AFM and it was summarized in Table 2. The surface roughness was 0.07 µm when introducing 100 sccm H_2S (samples 1 and 3). However, surface roughness increased to 0.13 µm as H_2S gas amount was increased to 700 sccm (samples 2 and 4). The high surface roughness of samples 2 and 4 was due to the porous layer formation with increase of the mixed gas.

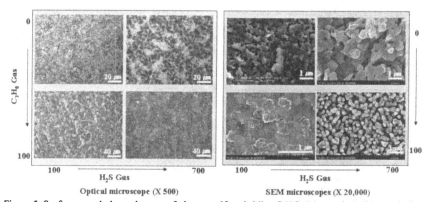

Figure 5. Surface morphology changes of plasma sulfur nitriding S45C; (a) sample 1, (b) sample 2, (c) sample 3, and (d) sample 4

Surface morphology changes with C_3H_8 gas addition during sulfur nitriding were observed by optical micrographs and SEM micrographs as shown in figure 5. The optical microscope showed that surface morphology were changed to the dense structures and darkened by addition of C source gas. This dense structure must be formed at the early of the sulfur nitriding by the interaction of C and S atoms and this resulted in thin compound layer of the samples 3 and 4. An interesting feature was observed by SEM micrographs of the C_3H_8 gas addition samples. With C source gas addition, their particle size was decreased to nano-size range less than 50 nm and this resulted in the dense microstructure as shown in the right side of figure 5. This dense and nano-sized surface microstructure of the samples 3 and 4 was main reason of the high surface hardness even the sample 4 had the porous microstructure. Another interesting feature was that an increase of carbon gas flow to 500 sccm was resulted in the formation amorphous-like compound layer.

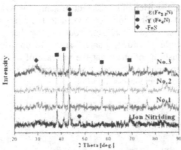

Figure 6. X-ray diffraction patterns of sulfur nitriding samples.

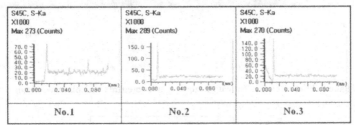

Figure 7. EPMA results of plasma sulfur nitriding sample

Phases of sulfur nitriding samples were analyzed with XRD and they are shown in figure 6. In the XRD results, $Fe_{2-3}N$ is dominant phase on the sulfur nitriding layer. It is not easy to detect FeS phase because their peaks were overlapped with those of $Fe_{2-3}N$ or Fe_4N phase. Only FeS peak observed in XRD analysis was detected at $2\theta=29°$. The existence of FeS phase in the sulfur nitriding samples could be detected by XPS analyses. To investigate the sulfur distribution near surface region, EMPA analyses were performed on the samples 1, 2 and 3 and the results are shown in figure 8. EPMA depth profile of S was observed up to 100 μm from the surface. As can be seen in figure 7, sulfur was detected up to about 10 μm from the surface. This indicated that decomposed sulfur from H_2S gas didn't diffuse to matrix but remained near surface. As H_2S gas amount increases (sample 2), sulfur concentration on sulfur nitriding surface increased as well. In addition, C_3H_8 gas addition (sample 3) also provided higher surface sulfur concentration of sample 3 compared with sample 1 of only H_2S mixed gas. From the analyses of XRD, EPMA, FeS was found to be formed on top surface regions during sulfur nitriding and FeS may have been related to porous structure of compound layer.

Figure 8 shows friction coefficient of various plasma sulfur nitriding samples with varying H_2S and C_3H_8 gas amount. Several researches have reported that friction coefficient was 0.8~1.0 for gas nitriding and 0.6~0.8 for plasma nitriding [10]. According to the results on the sulfur nitriding processes, the best friction coefficient data was 0.3 [6-9]. In the present study, friction coefficient was 0.4 and 0.5 for sample 1-3. For the plasma sulfur nitriding samples only with H_2S mixed gas, the best friction coefficient was obtained with the porous structured sample prepared with 700 sccm mixed gas as in the sample 2. With the addition of C_3H_8 gas, the dense and nano-size grained surface was formed as in the sample 3 but this structure was not good for the wear property. Its friction coefficient was as high as 0.45. However, the friction coefficient of sample 4 was significantly lowered to 0.2. Such low friction coefficient of sample 4 may be related to the nano sized porous structure layers. Since friction coefficient decreased with decreasing contact areas between two surfaces [11,12], the nano-sized porous layer of sample 4 reduced contact area and this resulted in the decrease of the friction coefficient. As previously mentioned that the increase of C_3H_8 gas amount up to 500 sccm was resulted in the formation amorphous like structure as shown in far right side of the figure 9, this structure had the lowest friction coefficient less than 0.2.

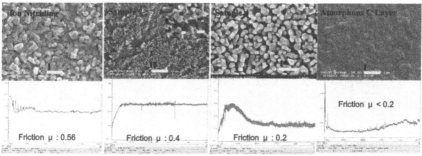

Figure 8. Friction coefficient with various samples prepared with plasma nitriding.

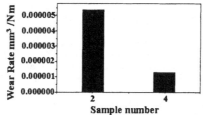

Figure 9. Wear rate of plasma sulfur nitriding S45C samples; (a) sample 2, (b) sample 4.

Wear rate of samples 1-4 was calculated from wear track and it is shown in figure 9. From the tribological tests, sample 4 treated under both H_2S and C_3H_8 gases had low friction coefficient and low wear rate from high hardness. The wear rate of sample 4 was 1×10^{-6} mm^3/Nm which is one order lower than plasma nitriding sample. According to the SEM observation on the wear track of sample 4 as shown in figure 10, porous layer structure was still remained after 1000 m of ball on disc test. Some pores were filled with materials that were worn out from the surface. This fine particle was also shown on the surface of the ball after wear test. They may prevent the direct contact of ball on the surface of the sample and this resulted in such low wear rate of samples treated in high H_2S and C_3H_8 gas. That kind of fine particles attached on the ball during wear test could act as a solid lubricant and this will be very effective to prevent the worn-out of the count part without the sulfur nitriding. The changes of the weight of the ball and the specimen after wear tests were measured and they are summarized in table II. Only plasma nitriding sample had low change in the weight of the specimen. The weight increase of the sample must result from the oxidation during wear test. The oxidation was confirmed by SEM analyses with EDS. But there was large decrease of the ball that is the counter part of the nitriding sample. This means that the surface of the sample prepared by nitriding would be protected from the wear but the counter part as the ball could be worn out easily by contact with high hardness surface. However, the worn rate could be decreased by the sulfur nitriding. Furthermore, the weight of the counter, the ball, increased in samples 2 and 4. It is thought that the fine and brittle microstructure was removed easily from the surface and attached to the ball and it could protect the both samples during the wear test. Fine and porous structure shown in sample 4 must be much effective for this process so that the weight decrease of the sample was distinctive and it was the same as the increase of the ball weight. That must be the reason for the best wear property of the sample 4.

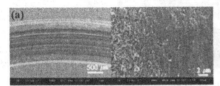

Figure. 10. SEM micrographs of plasma sulfur nitriding S45C after wear test;
(a) sample 2, (b) sample 4.

Table II. Weight changes of ball and the plasma sulfur nitriding S45C specimens after wear test.

Unit : g	Ball	Specimen	Total
Ion nitriding	- 0.0035	0.0003	0.0032
Sample 2	0.0005	0.0003	0.0008
Sample 4	0.001	- 0.001	0

There were changes of surface microstructure with plasma sulfur nitriding temperature. With the increase of the process temperature, the surface microstructure increased from nano-sized grain less than 50 nm to micro size over 300 nm, as shown in figure 11. According to the wear test, the best friction coefficient was obtained in the sample 4 that was prepared at 500 ° C with 700 sccm H_2S mixed gas and 100 sccm C_3H_8 gas. They had the fine and porous surface microstructure and this resulted in the best friction properties.

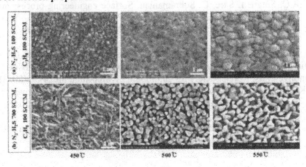

Figure 11. The change of surface morphology of the plsma sulfur nitriding samples with temperature

Figure 13. The applications of sulfur nitriding process

Figure 13 shows the possible industrial applications of the sulfur nitriding that would be useful for the wear resistance.

CONCLUSION

Properties of plasma sulfur nitriding S45C steel was examined with addition of H_2S/N_2 mixed gas and C_3H_8 gases. As the amount of H_2S gas increased, porous layer were formed on top of compound layer. In addition, sample treated under $C_3H_8 + H_2S$ gas mixture resulted in the nano-sized structure less than 50 nm. Unlike porous layers by conventional plasma sulfur nitriding samples, the nano sized porous layers had high hardness of about 800 Hv. Also, a fine distribution of nano and porous FeS particles improved the friction coefficient and maintained it less than 0.2. The wear rate was also reduced to 1 x 10^{-6} mm^2/Nm. Therefore, it is thought that C_3H_8 gas addition during plasma sulfur nitriding provides high hardness as well as low friction coefficient and good wear properties.

REFERENCES
[1]B. Edenhofer, Physical and metallurgical aspect of ion nitriding, Heat Treat Met., 2, 59-67 (1974).
[2]M.B. Karamis, Tribological behavior of plasma nitrided 722M24 material under dry sliding condition, Wear, 147, 385-99 (1991).
[3]K.N. Strafford, R.C. Smart, I. Sare, C. Subramanian. Surface engineering process and applications. New Holland Avence, USA: Technomic Publishing, p 157-70 (1995).
[4]E. Matin, O.T. Inal. Formation and growth of iron nitrides during ion-nitriding, J. Mater. Sci., 22, 2783-8 (1987).
[5]Y. Sun, T. Bell, Plasma surface engineering of low alloy steel, Mater. Sci. Eng. A, 140, 419-34 (1991).
[6]Ikmin Park, Chul Park, Insup Lee, Youncho Joun, Development of Plasma Sulf-Nitriding Technology for Die of Automotive Parts, Korea Society for automobile, symposium, pp.24-30 (2003).
[7]J.S. Lee, H.G. Kim, Y.Z. You, The Microstructures and Properties of Duplex Layer on the Tool steel Formed by Post-oxidation and Sulf-nitriding, J. of the Korean Society for Heat Treatment, 14, 81-8 (2001).
[8] J.S. Lee, D.K. Park, Y.Z. You, The Microstructures and Properties of Layer on the SCM440 steel formed by Plasma Sulf-nitriding. Korean Society for surface, 31, 266-277 (1998).

[9]Insup Lee, Ikmin Park. Solid lubrication coating of FeS layer on the surface of SKD61 steel produces by plasma sulf-nitriding, Surface & Coating Technology, **200**, 3540-3543 (2006).

[10]S.H. Choa, S.K. Kim, J.S. Park, Tribological Characteristics of Plasma Ion Nitriding Surface Treatment, Korea Society for lubrication, **12**, 60~70 (1996).

[11]'Friction and Wear of Meterials', Ernest Rabinowicz, John Wiley and sons, (1995).

[12]'TRIBOLOGY', I. M. HUTCHINGS, CRC Press. (1992).

ADHESION IMPROVEMENT OF HARD BORON NITRIDE FILMS BY INSERTION OF VARIOUS INTERLAYERS

Tetsutaro Ohori, Hiroki Asami, Jun Shirahata, Tsuneo Suzuki, Tadachika Nakayama,
Hisayuki Suematsu and Koichi Niihara

Extreme Energy-Density Research Institute, Nagaoka University of Technology
1603-1 Kamitomioka, Nagaoka, Niigata 940-2188, Japan

Mail: t_ohori@etigo.nagaokaut.ac.jp

ABSTRACT

Significant reduction of compressive stress of cubic boron nitride (c-BN) thin film and substantially improved adhesion was achieved by a new coating concept consisting of two steps; adding He gas in the thin film preparation process and inserting interlayers. The c-BN film with a thickness of 600 nm was deposited by radio frequency magnetron sputtering using a pure B target on silicon (100) substrate with various interlayers. It was found that the B layer exhibited a good adhesion. The present study was focused on the investigation of the morphology, the macrostructure using optical microscopy, microstructure of cross-section scanning electron microscopy (SEM), and chemical bonding of the thin films by Fourier transform infrared spectroscopy (FT-IR).

INTRODUCTION

Many attractive properties in cubic boron nitride (c-BN) make it an excellent candidate material for numerous industrial applications. Desirable characteristics for the industrial applications are such as the second highest hardness next to diamond, chemical inertness against ferrous-based metals and oxidation resistance at high temperatures[1], large thermal conductivity and widest optical band gap among III-V semiconductors[2]. However, c-BN thin films which can be used in industries have not been produced, since c-BN thin films do not have enough adhesion strength to the substrates. Reasons of this poor adhesion strength are attributed in part to growth of amorphous and sp^2 bonded boron nitride initially formed on a substrate, and to form large residual compressive stress in c-BN thin films. These facts are the reasons for the thin films peeling off from the substrates.

By most coating techniques, available c-BN thin films without peeling off were limited within thickness of 100-300 nm. Presently, only eight groups have reported, to our knowledge, on relatively thick c-BN coatings in micrometer-scale[2-9]. Especially, the following three groups have reported a lot of articles. Matsumoto et al. succeeded in preparation of c-BN thin films with thickness of 20 μm by DC jet plasma chemical vapor deposition (CVD)[6, 10], however, very high temperatures up to 1000 °C were necessary. Bewilogua et al. were able to grow 2 μm thin films using a B₄C target for inserting an interlayer by a sputtering technique[7, 11, 12]. Ulrich et al. reported 2~3 μm thick c-BN films using additional O₂ gas[9, 13]. These studies were needed 3rd element to improve adhesion strength and/or to released internal stress. Therefore, it is still a challenge to develop new deposition methods for thick c-BN films.

Our study is based on the deposition of a two layered coating consisting of an interlayer and a top layer of c-BN. The c-BN layer was made from a B target and N_2 gas. Additional He gas was used for inert gas to reduce the residual stress. Moreover, in order to get higher adhesion strength, various interlayers were tried to use such as B, B₄C, TiN, AlN, CrN and WC.

EXPERIMENTAL SETUP

The growth of the BN thin films has been performed by the radio frequency (RF) magnetron sputtering technique with a substrate rotation mechanism. Figure 1((a) shows the schematic of the experimental setup used in this study. Pure B was used as a large size target (88mm in wide and 200mm in length square shape disk, 99.0 up% purity) for deposition of the top c-BN layer. Similar size targets were used for the deposition with various interlayers. A single crystalline silicon (100) wafer was cut into pieces with size of $25 \times 25mm^2$, which were used as substrates. Six substrates were set linearly on a substrate holder like Fig. 1((b). Substrate temperature was kept at about 450 °C, and base pressure was lower th an 1×10^{-3} Pa by using a rotary pump and a turbo molecular pump.

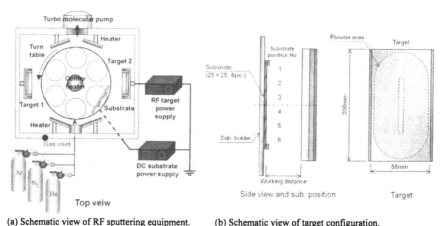

(a) Schematic view of RF sputtering equipment.   (b) Schematic view of target configuration.

Figure 1. Schematic of experimental equipment.

Before the thin film deposition, the substrates were cleaned by sputter etching with introduction of pure argon through the gas inlet at 400 sccm with -650 V substrate bias for 20 minutes. A DC substrate power supply was used and operated with pulsed mode at 250 kHz.

The B target was used for the preparation of the top layer (target 2 in Fig. 1 ((a)). During the top layer deposition, an RF target power supply was operated at the power of 500 W and frequency of 13.56 MHz. The DC voltage of -100 V was supplied to the substrate. Working gases were Ar, N_2 and He, each flow rate were controlled as follows; Ar 50 sccm, N_2 20 sccm and He 45 sccm. The deposition time was 480 minute. Typical processing parameters for the interlayer and the BN film preparation are listed in Table I.

The targets for the deposition of interlayers were B, B_4C, Ti, Al, Cr and WC (target 1 in Fig. 1 (a)). During the deposition of interlayers, a target power supply was operated at the power of 500 W and frequency of 13.56 MHz. A DC voltage of -100 V was supplied to the substrate. Working gases and the flow rate are listed in Table II.

In this work, chemical bonding of the thin films was investigated by Fourier transform infrared spectroscopy (FT-IR: JASCO, FT/IR-4200). Furthermore, the raw data were typically recorded for 50 scans from 4000 cm^{-1} to 500 cm^{-1} with the resolution of 4 cm^{-1} and were measured by transmission through both BN thin films and the substrate. The absorption spectra were obtained by subtracting the Si substrate background intensity from the raw data. All samples were measured within one day after the deposition. The absorption bands at 1380cm^{-1} and 780 cm^{-1} are due to in-plane and out-of-plane vibration mode of hexagonal boron nitride (h-BN)[14], and around 1080 cm^{-1} are due to residual spectrum of c-BN[15]. c-BN and h-BN intensities of FT-IR spectra were described as I_{cBN} and I_{hBN}, respectively. These were normalized from FT-IR spectra intensities of the absorbance at approximately 1050-1100cm^{-1} and 1380cm^{-1}, c-BN fraction (F_{cBN}) was defined as follows[16],

$$F_{cBN} = \frac{I_{cBN}}{I_{cBN} + I_{hBN}} .$$ (1)

The macroscopic morphology observation of the films was carried out with an optical microscope (Keyence, VHX-900). The microstructural observation of the film surface and cross section were carried out with a scanning electron microscope (SEM: JEOL, JSM6700F). The crystalline structure of the thin films was characterized by X-ray diffraction (XRD: Rigaku, RINT 2500HF+/P(C) technique using Cu-Kα radiation (λ = 1.5418 Å) with an operating voltage of 50 kV and a current of 300 mA. XRD patterns of the samples were obtained in 2θ between 20° to 60° with a step of 0.02° and a scan speed of 2°/min.

Table I. Parameters for deposition of BN films

	Pre-sputter process	Interlayer process	Top layer process
Substrate		Si(100)	
Base pressure [Pa]		$< 1 \times 10^{-3}$	
Table rotation [rpm]		1.0	
Substrate temperature [°C]		~ 450	
Target material	-		B (99 up%)
Time [min]	20		480
Sputtering method	-		RF(13.56MHz)
Cathode power [W]	-	Table II	500
Sub. bias voltage [V]	-650		-100
Ar gas flow rate [sccm]	400		50
N₂ gas flow rate [sccm]	-		20
Additional element		He gas	
Additional gas flow rate [sccm]		He:45	

Table II. Parameters for the interlayer process

	B	B₄C	WC	TiN	CrN	AlN
Target material	B	B₄C	WC	Ti	Cr	Al
Time [min]	120	60	60	60	60	60
Sputtering method	RF	RF	RF	DC	RF	RF
Cathode power [W]				500		
Sub. bias voltage [V]				-100		
Ar gas flow rate [sccm]	100	100	400	400	400	400
N₂ gas flow rate [sccm]	-	-	-	100	100	100

RESULTS AND DISCUSSION

1. Changed surface condition by He gas addition

In this section, He addition effect in c-BN thin films deposition using B target is described. FT-IR spectra of the deposited thin films with and without additional He gas are shown in Fig. 2. In Fig. 2(a) and (b) within the range of wave numbers between 500 cm⁻¹ and 2000 cm⁻¹, a strong absorption peak due to c-BN was observed at about 1080 cm⁻¹. In Fig. 2(b), the absorption bands at 1380 cm⁻¹ and 780 cm⁻¹ due to an h-BN phase were also observed. The film deposited without He gas addition (Fig. 2(a)) contained about 90% c-BN fraction calculated from the formula (1). The c-BN fraction of the thin film deposited with additional He gas, (Fig. 2(b)) was decreased to 72%.

Plan view optical microscope images of the deposited thin films with and without additional He gas are shown in Fig. 3(a) and (b) and magnified images by SEM are shown in Fig. 4(a) and (b). From Figs. 3(a) and 4(a), it was found that the thin film deposited without He gas was peeled off and became small spalls. The reason for this peeling off is large residual compressive stress in the film[17]. It is considered that bending stress originated from the compressive stress spalled the peeled thin film. High c-BN fraction samples deposited with only Ar gas showed peeling off like spalled shape as same as the sample of Figs. 3(a) and 4(a). On the other hand, although the sample deposited with He gas was peeled off as shown in Fig. 3(b), the film morphology was different from that of the sample deposited without He, and the peeled thin film was shriveled and not spalled. The magnified view of the sample deposited with additional He gas was shown in Fig. 4(b). A crack was observed on some views and small spalls such as shown in Fig. 4(a) were not observed.

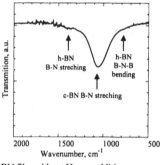

(a) BN film without He gas addition.

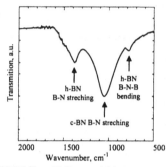

(b) BN film with He gas addition.

Figure 2. FT-IR spectra of BN films.

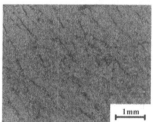

(a) BN film surface without He gas addition.

(b) BN film surface with He gas addition.

Figure 3. Macroscopic image of BN films by digital microscope.

(a) BN film surface without He gas addition.

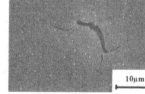

(b) BN film surface with He gas addition.

Figure 4. Magnified view image by SEM of BN films.

The reasons of decreasing c-BN fraction and changing cracking mode with and without He addition are not clearly known. It seemed that the redial stress was released by additional He gas. Other reasons including surface roughness, young's modulus change and the microstructural change can also suppress the cracking. The fact remains as a matter to be discussed further.

2. Selection from various interlayers during the Si substrate and the BN film

It has been shown that the film condition was improved by the additional He gas. However, the adhesion strength was not sufficient as described above. In this section, effect of inserting interlayer between substrate and c-BN film was described. Table III summarizes the results of film thickness of various interlayers. Surface morphology of BN films with various interlayers are shown in Figs. 5(a) ~ (c). As shown in Fig. 5(a), the BN film with a CrN interlayer was peeled off as same as the case shown in Fig. 3(b). CrN interlayer did not improve adhesion strength. Morphology of the BN films with interlayer of WC, Ti, and AlN were similer as that in Fig. 5(a). Figure 5(b) shows the BN films with a B_4C interlayer has the small area peeled off. It seemed that the B_4C interlayer improved some adhesion strength. Figure 5(c) shows a B interlayer have good adhesion between the Si substrate and the BN film. In our result, B is the best interlayer from the view point of peeling off.

Table III. Result of interlayer

	B	B_4C	WC	TiN	CrN	AlN
Deposition time [min]	120	60	60	60	60	60
Interlayer thickness [nm] (Measured from SEM image)	80	60	few	30	300	110

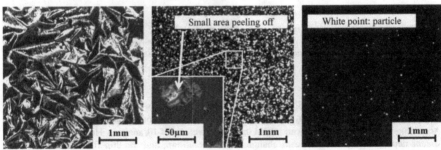

(a) BN film with inserting CrN interlayer.

(b) BN film with inserting B_4C interlayer.

(c) BN film with inserting B interlayer.

Figure 5. Macroscopic image of BN films with interlayers of (a) CrN, (b) B_4C, and (c) B.

3. Deposition of thick film using B interlayer

In the last section, the increase in adhesion strength of thick c-BN films with inserting a B interlayer is described. Deposition times of top layer were 20 hours in only this experiment. The formation of c-BN phase in the deposited films is evidenced by the absorption peak of FT-IR as shown in Fig. 6. c-BN fraction was about 74%. Figure 7 show a cross-section SEM image of c-BN film. It was concluded that we succeeded to deposit a c-BN film of total 680 nm thickness, including a 600 nm c-BN top layer and an 80 nm B interlayer. It was not possible to deposit BN thin films with thickness of 600 nm either with interlayers or in He gas.

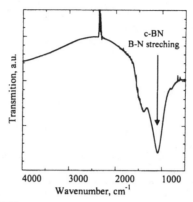

Figure 6. FT-IR spectra of thick c-BN films with boron interlayer.

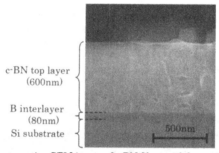

Figure 7. Cross-section SEM image of c-BN films with boron interlayer.

CONCLUSION

The growth of c-BN films by means of radio frequency magnetron sputtering has been achieved by He gas addition into the deposition process in combination with inserting B interlayer. According to optical microscopy and SEM images as well as FT-IR analysis, He gas addition during sputter deposition led to significant change in film condition, and the thin films with high cubic phase fraction were obtained without film peeling off by an inserting B interlayer. The maximum 600 nm thickness was obtained for a thin film deposition with non reactive atoms except for nitrogen.

REFERENCES

[1] L. Vel, G. Demazeau and J. Etourneau, Cubic boron nitride: synthesis, physicochemical properties and applications, Materials Science and Engineering, **B10**, 149-164 (1991).

[2] D. Litvinov, C. A. Taylor II and R. Clarke, Semiconducting cubic boron nitride, Diamond and Related Materials, **7**, 360-364 (1998).

[3] K. -L. Barth, A. Lunk and J. Ulmer, Influence of the deposition parameters on boron nitride growth mechanisms in a hollow cathode arc evaporation device, Surface and Coatings Technology, **92**, 96-103 (1997).

[4] K. -B. Kim and S. -H. Kim, Adhesion improvement of cubic BN:C film synthesized by a helicon wave plasma chemical vapor deposition process, Journal of Vacuum Science and Technology, **A18**, 900-906 (2000).

[5] H. -G. Boyen, P. Widmayer, D. Schwertberger, N. Deyneka and P. Ziemann, Sequential ion-induced stress relaxation and growth: A way to prepare stress-relieved thick films of cubic boron nitride, Applied Physics Letter, **76**, 709-711 (2000).

[6] S. Matsumoto and W. Zhang, High-rate deposition of high-quality, thick cubic boron nitride films by bias-assisted DC jet plasma chemical vapor deposition, Japanese Journal of Applied Physics, Part 2: Letters, **39**, 442-444 (2000).

[7] K. Yamamoto, M. Keunecke, K. Bewilogua, Zs. Czigany and L. Hultman, Structural features of thick c-boron nitride coatings deposited via a graded B–C–N interlayer, Surface and Coatings Technology, **142-144**, 881-888 (2001).

[8] G. Bejarano, J.M. Caicedo, E. Baca, P. Prieto, A.G. Balogh and S. Enders, Deposition of B₄C/BCN/c-BN multilayered thin films by r.f. magnetron sputtering, Thin Solid Films, **494**, 53-57 (2006).

[9] S. Ulrich, E. Nold, K. Sell, M. Stüber, J. Ye and C. Ziebert, Constitution of thick oxygen-containing cubic boron nitride films, Surface and Coatings Technology, **200**, 6465-6468 (2006).

[10] S. Matsumoto and W. J. Zhang, The introducing of fluorine into the deposition of BN: a successful method to obtain high-quality, thick cBN films with low residual stress, Diamond and Related Materials, **10**, 1868-1874 (2001).

[11] K. Bewilogua, M. Keunecke, K. Weigel and E. Wiemann, Growth and characterization of thick cBN coatings on silicon and tool substrates, Thin Solid Films, **469-470**, 86-91 (2004).

[12] M. Keunecke, E. Wiemann, K. Weigel, S.T. Park and K. Bewilogua, Thick c-BN coatings – Preparation, properties and application tests, Thin Solid Films, **515**, 967-972 (2006).

[13] M. Lattemann, S. Ulrich and J. Ye, New approach in depositing thick, layered cubic boron nitride coatings by oxygen addition—structural and compositional analysis, Thin Solid Films, **515**, 1058-1062(2006).

[14] R. Geick, C. H. Perry and G. Rupprecht, Normal modes in hexagonal boron nitride, Physical Review, **146**, 543-547 (1966).

[15] P. J. Gielisse, S. S. Mitra, J. N. Plendl, R. D. Griffis, L. C. Mansur, R. Marshall, and E. A. Pascoe, Lattice infrared spectra of boron nitride and boron monophosphide, Physical Review, **155**, 1039-1046 (1967).

[16] P. B. Mirkarimi, K.F. McCarty and D.L. Medlin, Review of advances in cubic boron nitride film synthesis, Materials Science and Engineering: R: Reports, **21**, 47-100 (1997).

[17] J. Hahn, M. Friedrich, R. Pintaske, M. Schaller, N. Kahl, D.R.T. Zahn and F. Richter, Cubic boron nitride films by d.c. and r.f. magnetron sputtering: layer characterization and process diagnostics, Diamond and Related Materials, **5**, 1103-1112 (1996).

PRODUCTION OF ALUMINA MATRIX NANOCOMPOSITE BY SOLID STATE PRECIPITATION

Amartya Mukhopadhyay and R. I. Todd
Department of Materials, University of Oxford, Oxford OX1 3PH, UK

* Corresponding author, Tel: +44 1865 273718; Fax: +44 1865 273783; E-mail: richard.todd@materials.ox.ac.uk

ABSTRACT

The present communication reports a novel processing route for development of alumina-based nanocomposites by precipitating oxide nanoparticles during aging of alumina solid solutions. Al_2O_3 containing 10 wt.% Fe_2O_3 (haematite) was first sintered in air at 1450°C resulting in complete dissolution of Fe_2O_3 in Al_2O_3 (solution treatment). Aging of the fully dense polycrystalline solid solutions in a reducing atmosphere of N_2 + 4% H_2 at 1450°C for different durations (up to 50 hrs) led to the precipitation of up to ~ 20 vol.% $FeAl_2O_4$ particles. The influence of aging duration on the microstructure development (amount, size and distribution of secondary phase particles) and mechanical properties (Vickers hardness and indentation toughness) has been studied in detail. After aging for ~ 10 hrs nano sized precipitate particles (~ 80 nm) appeared within the matrix grains, while coarser micron sized particles were present on the grain boundary regions. This resulted in the development of 'hybrid nano/micro ceramic composite' via reduction aging using cheaper processing technique. The nanocomposites developed on reduction aging for time intervals between 10 - 20 hrs possessed considerably improved indentation fracture toughness (by ~ 45%) with respect to pure monolithic Al_2O_3. However, on aging for a more extended period of 50 hrs, significant coarsening of the secondary phase particles was observed accompanied by reduction in fracture toughness and hardness.

1. INTRODUCTION

With the recognition that ceramic nanocomposites possess some appealing mechanical, physical and tribological properties, they have been the subject of considerable research activities in recent years [1-11]. However, to date, most of the developmental research on ceramic nanocomposites has been focused on systems comprising at least one non-oxide component [1-11]. Though such materials have been reported to possess beneficial high temperature properties, yet practical applications of such nanocomposites in air may leads to problems related to oxidation of the non oxide phases. Furthermore, it has been realised that reaction between the components and formation of deleterious reaction products either during sintering or during high temperature application might limit their practical applicability. In addition, processing of such nanomaterials usually require advanced sintering techniques like hot pressing or spark plasma sintering which are not favourable for mass scale commercial production. Hence, it is imperative to develop ceramic nanocomposites consisting of only oxide phases that can be produced using cheaper and more conventional processing techniques so that these promising materials can be used commercially.

One of the technologically important structural ceramics, which is widely used for various structural and wear resistance applications, is alumina (Al_2O_3). Despite possessing very high melting point, Young's modulus and hardness, along with considerable wear resistance, the major drawbacks of monolithic alumina, limiting its structural applications, are moderate strength (~ 350 MPa) and fracture toughness (~ 3 MPa $m^{1/2}$). This issue was initially addressed by Niihara in his pioneering paper concerning the development of ceramic nanocomposites such as Al_2O_3-SiC [1]. The incorporation of nanocrystalline SiC in an alumina matrix has been found to result in improvement of strength (by ~ 20 %) and wear resistance (by factor of ~ 3) [3,4,7-9]. In addition to the extensive research on Al_2O_3 - SiC nanocomposites, reports are also available on the successful development of other nanocomposites based on alumina, such as Al_2O_3 - TiC [10], Al_2O_3 - TiCN - SiC [11], with all showing improved properties with respect to monolithic alumina. However, the presence of such refractory non oxide particles renders processing difficult via conventional pressureless sintering techniques and also limits the high temperature applicability of such materials. The development of oxide/oxide nanocomposites is obviously desirable but is difficult because most oxides are less refractory than SiC and other covalent ceramics and coarsening during sintering therefore prevents the retention of nano scaled oxide dispersions when processed using conventional powder metallurgical routes [4].

One of the possible routes to develop ceramic nanocomposites is in-situ formation of nano-sized dispersoids by precipitation during aging of supersaturated solid solutions. However, to date there are very few reports which explore the possibility of development of oxide-oxide ceramic nanocomposites via such precipitation techniques. One of the earliest indications of precipitation from supersaturated solid solution in ceramics was the needle shaped precipitates (possibly TiO_2) observed in star sapphire [12,13]. In a more recent investigation, Wang et al. [6] developed Al_2O_3-based nanocomposites, possessing improved mechanical properties, by precipitation of nanocrystalline magnesium aluminium spinel during aging in a reducing atmosphere of a supersaturated solid solution of Al_2O_3 containing Mg^{2+} and Ti^{4+}. However, due to limited solubility in Al_2O_3 only a maximum of 2 mol% $(MgO + TiO_2)$ could be incorporated during solution treatment which severely limited the volume fraction of the secondary phase particles which could be precipitated out during the aging treatment.

The present work reports preliminary results aiming to develop oxide/oxide nanocomposites based on Al_2O_3 using cheaper conventional processing route (pressureless sintering), via aging technique. One of the major problems associated with this process stems from the fact that most cations (as oxides) present negligible solubility in Al_2O_3 due to their comparatively larger radii and different crystal structure of their oxides compared to that of α-Al_2O_3. Fortunately, Fe_2O_3 (haematite), belonging to the same space group ($R\bar{3}c$) as α-Al_2O_3 (corundum), possesses substantial solubility in Al_2O_3 at moderately high temperatures in air (~ 15 wt.% at ~ 1410°C) [14]. Furthermore, under reducing conditions Fe^{3+} (of dissolved Fe_2O_3) can be reduced to Fe^{2+}, which being aliovalent to Al^{3+} (of Al_2O_3) possess negligible solubility in Al_2O_3. Hence, aging of 'supersaturated' Al_2O_3 - Fe_2O_3 solid solutions under controlled reducing conditions should lead to the formation of oxide precipitate particles containing Fe^{2+}.

2. EXPERIMENTAL PROCEDURE

2.1 Processing

In the present investigation, high purity (~ 99.995%), ultra-fine (particle size ~ 200 nm) α-Al_2O_3 powders (AKP50, Sumitomo, Japan) were used as the starting material. The Al_2O_3 powders, doped with 250 ppm of pure MgO (99.95 %; 120 nm) powders, were dispersed in ethanol using an ultrasonic probe. The MgO doping helps in preventing abnormal grain growth in Al_2O_3 and hence achieve near theoretical density on sintering. Incorporation of Fe^{3+} was achieved using $Fe(NO_3)_3.9H_2O$ (Sigma Aldrich, UK, purity > 98%), which is soluble in ethanol and thus achieved a homogeneous distribution of Fe^{3+} in the alumina slurries. The $Fe(NO_3)_3.9H_2O$ solutions contained requisite amount of the salt, which would result in doping levels corresponding to 10 wt.% Fe_2O_3 in Al_2O_3. The pure alumina slurries, as well as the slurries containing $Fe(NO_3)_3.9H_2O$ solutions, were ball milled for 24 hours in bottles made of polyethylene, using high purity (99.99%) alumina balls. During drying on a hot plate, the slurries were constantly stirred, using a magnetic stirrer, to prevent segregation. The dried powders were lightly crushed to remove agglomerates and softly ground in a mortar and pestle and passed through a 150 μm sieve.

Green compacts were produced by uniaxial pressing (at 100 MPa) of the powders into 20 mm discs. The green compacts were pressureless sintered in air inside the alumina tube of a high temperature tube furnace (Lenton, UK). The samples containing 10 wt.% Fe_2O_3 will henceforth be referred to as 'A10F'. During the sintering cycle, the A10F green pellets were slowly heated (at 5°C/min) up to a temperature of 750°C and held there for 1 hour in order to decompose the ferric nitrate. Following this, a heating rate of 10°C/min was used up to the sintering temperature of 1450°C. The pure Al_2O_3 pellets were directly heated to the sintering temperature of 1450°C from room temperature at a heating rate of 10°C/min. The selected sintering temperature of 1450°C corresponds to the temperature on the Al_2O_3 - Fe_2O_3 phase diagram, where ~ 15 wt.% Fe_2O_3 is soluble in Al_2O_3. Hence it is expected that the maximum possible amount of Fe_2O_3 generated (~ 10 wt.% in A10F) on complete dissolution of ferric nitrate, will be dissolved in Al_2O_3 during sintering, thus yielding an alumina-based solid solution containing Fe^{3+} cations.

For ensuring complete dissolution and homogenous distribution of Fe^{3+} in the cation sub-lattice of Al_2O_3, the samples were held at the sintering temperature (1450°C) for 5 hrs (solution treatment) [6]. After completion of the solution treatment, in order to avoid possible precipitation of secondary phase during furnace cooling, the samples were 'quenched' by using an alumina rod to push them immediately to the end of the furnace tube, where the temperature was ~ 100°C. In order to develop nanocomposites based on alumina, the supersaturated solid solutions were aged under reducing conditions using a N_2 - 4% H_2 (forming) gas mixture at the temperature of 1450°C for different time intervals up to 50 hrs.

2.2 Characterisation
The sintered densities of samples were measured in distilled water according to Archimedes' principle. For phase evolution and microstructural characterization, all the samples were ground and polished using 25, 6, 3, 1 and 0.25 μm diamond slurries in order to remove the external surfaces and produce optically reflective ceramographic surfaces representative of the sample cores. The phase identification of the green pellets, sintered samples and aged samples was performed by X-ray diffraction (XRD) using CuKα radiation (Philips, Japan) at collection rate of 3°/min. In order to reveal the grain boundaries of the as sintered pure Al_2O_3 and A10F solid solution samples, thermal etching was done at a temperature of 1400°C for 15 min in air. The microstructures of all the polished surfaces were observed using field emission scanning electron microscopy (JEOL 6500F) operated at 20 KV. Elemental analysis of the microstructural phases was performed using EDS, attached to the SEM. Volume fractions of the secondary phase particles were determined via point counting on back scattered electron (BSE) images and the average particle sizes were determined from the mean of the major and minor axes of at least 20 particles, multiplied by 1.27 to obtain equivalent circular diameters. Sample preparation for transmission electron microscope (TEM) observations involved mechanical thinning of 3 mm discs, followed by ion-milling to electron transparency in a Gatan Duo Mill (5 kV). Observations were performed using a 200 kV JEOL 2000FX TEM, equipped with an EDS system.
The hardness was measured by Vickers indentation (H_v) on polished surfaces with an indentation load of 2 kg and loading time 15s. For determination of indentation fracture toughness (K_{IC}), 'short crack' method, involving calculation of K_{1c} from measured crack lengths emanating from corners of indent diagonals, is used and such calculations are based on relations proposed by Anstis et al. [15]. Although absolute values of fracture toughness of brittle materials (like ceramics) tend to be more reliable if measured using 'long crack' methods (SENB, SEVNB) [15-21], the ranking of ceramics based on their relative values of fracture toughness can be reproducibly achieved using short crack methods [15-21].

3. RESULTS AND DISCUSSION

3.1. Phase evolution

XRD patterns, recorded with as sintered pure Al_2O_3, calcined A10F pellets as well as solution treated (as sintered) and aged (1450°C; 10 hrs; N_2 + 4% H_2 atmosphere) A10F are presented in Fig. 1. From Fig. 1a it can be observed that on calcination at 750°C for 1 hr α-Fe_2O_3, formed due to the decomposition of the initially added $Fe(NO_3)_3.9H_2O$, is present along with α-Al_2O_3. However, in the XRD pattern corresponding to the as sintered A10F no peak corresponding to α-Fe_2O_3 can be detected, which indicates that Fe_2O_3 (~ 10 wt.%) gets completely dissolved in the Al_2O_3 matrix during holding at the sintering temperature of 1450°C, in accordance with the Al_2O_3 - Fe_2O_3 phase diagram [14]. It has been critically observed that there is a notable shift in the peak position towards the lower angle (2θ) side on solution treatment of A10F. Since the ionic radius of Fe^{3+} (0.065 nm) is comparatively larger than that of Al^{3+} (0.054 nm), dissolution of Fe^{3+} in the Al_2O_3 matrix, via substituting for the Al^{3+} in the cationic sub-lattice, results in the expansion of the lattice and hence corresponding shifts in peak positions towards lower Bragg angles.

The most important result obtained from the XRD investigations has been the appearance of peaks corresponding to $FeAl_2O_4$ (iron aluminate spinel) on aging of solution treated A10F at 1450°C in reducing atmosphere (N_2 + H_2 gas mixture) (Fig. 1). Additionally, peaks corresponding to α-Al_2O_3 get shifted back towards higher Bragg angles. This is consistent with the decrease in lattice parameter due to diminution of the Fe present in solid solution. Hence, the XRD results reveal that aging of the supersaturated solid solution of A10F under reducing conditions results in the precipitation of $FeAl_2O_4$ as the secondary phase.

3.2. Microstructural development

3.2.1. As sintered microstructures

Fig. 2 presents representative back scattered scanning electron micrographs (BSE) of polished and thermally etched surfaces of pure Al_2O_3 (a) and A10F (b). The absence of significant residual porosity, as observed from the micrographs of all the sintered samples, reveals that good densification was achieved on sintering at 1450°C for 5 hrs. In fact, as determined using Archimedes' principle, all the samples were densified to > 98% of their theoretical density. For pure Al_2O_3 the grains are fairly equiaxed in shape, without any observable abnormal or anisotropic grain growth (Fig. 2a). The grain sizes vary between 2 - 4 μm. In contrast, the Fe containing samples show evidence of anisotropic grain growth. The microstructure is characterised by the presence of elongated grains in a matrix of finer equiaxed grains. The equiaxed grains possess sizes ranging between 1 - 3 μm, while some of the elongated grains have lengths varying between 5 - 8 μm and widths between 2 - 3 μm (Fig. 2b).

The distribution of the various elements concerned in the microstructure of as sintered A10F has been revealed by X-ray mapping (Fig. 3). It can be observed that Fe appears to be distributed uniformly throughout the microstructure. Tartaj and Messing [22] have previously reported that there is negligible grain boundary segregation of Fe^{3+} in this system owing to its higher bulk solubility in Al_2O_3. Hence Fe forms a homogeneous solid solution in Al_2O_3 during sintering at 1450°C.

3.2.2. Microstructural development on aging

As against the single phase (Al_2O_3 - Fe_2O_3 solid solution) microstructures observed for the as sintered (solution treated) A10F (Figs. 2 and 3), BSE images clearly reveal the appearance of a secondary phase on aging of the solution treated A10F for various durations at 1450°C in reducing atmosphere of N_2 + 4% H_2 (Fig. 4). The secondary phase appears brighter in contrast with respect to the matrix and EDX patterns obtained from such phase indicate that the secondary phase is richer in iron (having higher atomic mass) (Fig. 4c). Hence correlation with the XRD analysis, obtained from the same ground and polished surfaces (sample core), indicates that the brighter secondary phase particles appearing on aging are $FeAl_2O_4$ spinel particles.

The effect of aging duration (1450°C; reducing atmosphere) on the microstructural development, including the distribution and relative sizes of the secondary phase particles ($FeAl_2O_4$), can be observed from the BSE micrographs corresponding to the different aging times (0-50 h), as presented sequentially in Fig. 4 (a-d). Furthermore, the mean particle sizes, along with their locations with respect to the matrix (Al_2O_3 - Fe_2O_3 solid solution) grains are mentioned in Table 1. Only micron sized secondary phase particles (~ 1 μm) located along the matrix grain boundaries (intergranular) can be observed on aging at 1450°C without any holding time (Fig. 4a). Similar observations have been made on aging for a duration of 5 hrs. Such absence of any particles within the matrix grains (intragranular) can be correlated with the lower bulk diffusion coefficients which results in a comparatively larger incubation period for nucleation and growth of intragranular secondary phase particles even at a relatively high temperature of 1450°C. However, on aging for a duration of 10 hrs at 1450°C, extremely fine 'nano sized' secondary phase particles (~ 70 nm) can be observed within the matrix grains, along with the presence of coarser 'micron sized' particles (~ 1.7 μm) along the grain boundaries (Fig. 4b). Hence, aging for a certain duration of time under reducing conditions leads to microstructural development which can be classified as a 'hybrid nano/micro composite' microstructure. On continued aging up to 20 hrs, modest coarsening of the inter- as well as intra-granular particles (Fig. 4c and Table 1) occurs. The distribution of the secondary phase particles can be more clearly observed from the bright field transmission electron micrograph (TEM) corresponding to A10F aged for 20 hrs at 1450°C, which shows the presence of micron sized intergranular particle along with nano sized intragranular particles (< 200 nm) (Fig. 5). On aging for an extended duration of 50 hrs considerable microstructural coarsening occurs, with the intragranular particles growing to submicron size (~ 0.4 μm) along with the formation of a nearly interconnected network of intergranular spinel phase (Fig. 4d). It can be critically observed that in the composites containing intragranular particles there are precipitation free zones near the grain boundary (intergranular) particles. Such regions usually form during aging due to the depletion of the solute atoms which join the heterogeneously nucleated intergranular precipitate particles.

Percentage of volume occupied by the secondary phase in the Al_2O_3 - $FeAl_2O_4$ nanocomposites, as determined from BSE images via point counting method, are plotted as a function of aging duration in Fig. 6. Volume percent of ~ 20 % is achievable on aging at 1450°C for a period of 15 hrs or more. It can be further observed that, while the volume fraction of $FeAl_2O_4$ increases drastically due to Fe coming out of solid solution as a result of reduction aging up to ~ 15 hrs, the increment is much less pronounced on further aging. In correlation with the microstructural observations (Fig. 4 and 5), coarsening takes precedence over new particle formation during the latter half of the aging cycle (after ~ 15 hrs).

3.3. Mechanical properties

The room temperature mechanical properties (H_v and K_{1c}), as obtained via Vickers indentation (load: 2 kg), are reported in Table 1. It can be observed that in general the hardness values for the 'nanocomposites' vary between 19-20 GPa, which are nearly comparable to that of pure Al_2O_3 (~ 20 GPa). On the contrary, the hardness of the 'composite' (~ 18 GPa), containing coarser micron sized inter/intragranular secondary phase particles, developed by aging for an extended period of 50 hrs appears to be slightly less than that of pure Al_2O_3 and the 'nanocomposites' developed by aging for shorter duration (Fig. 4d and Table 1). However, consideration should also be given to the large scatter in the hardness values measured for the 'composite', which makes it difficult to specifically state that the hardness decreases with increase in aging time.

An important result of the present investigation is the possibility of improvement of indentation 'fracture' toughness of Al_2O_3 by development of such 'hybrid nano/micro composites' via precipitation of $FeAl_2O_4$ as the secondary phase particles during reduction aging treatment. It can be observed from Table 1 that on aging of the A10F solid solutions for a duration of 20 hrs, nearly 45% improvement of fracture toughness can be achieved.

Critical assessment of the indentation toughness, as reported in Table 1, points towards some relationship between the aging duration, microstructural development (volume fraction, size and distribution of the secondary phase particles) and the indentation toughness for the as developed composites. It can be observed that precipitation of a limited amount of micron sized $FeAl_2O_4$ particles along the matrix grain boundaries (aging durations: 0 and 5 hrs) does not result in any apparent improvement in the indentation toughness with respect to pure Al_2O_3 (3.2 MPa m$^{1/2}$). Aging for longer periods (10-20 hrs) results in precipitation of nano sized secondary phase particles within the matrix grains, and concomitantly considerable improvement can be noted for the indentation toughness, which reaches its peak (~ 4.6 MPa m$^{1/2}$) for the nanocomposite developed by reduction aging for 20 hrs at 1450°C. It can be observed from the secondary electron SEM images of the Vickers indents obtained at a load of 2 kg that although radial cracking is pronounced for pure Al_2O_3 (Fig. 7a), such cracking is comparatively suppressed for the 'nanocomposite' developed by reduction aging of A10F for 20 hrs at 1450°C (Fig. 7b). Another important observation which can be made from Fig. 7 is that, on being polished under similar conditions and for the same time, the 'nanocomposite' develops a much better surface finish due to a significantly lower degree of pull out via brittle fracture, with respect to that of pure Al_2O_3. However, on continued aging for a much longer duration of 50 hrs significant coarsening of the inter and intragranular $FeAl_2O_4$ particles occurred without any notable increase in their volume fractions, and this was accompanied by deterioration in the indentation toughness (~ 3.5 MPa m$^{1/2}$).

Observation of the fractured surfaces of as sintered pure Al_2O_3 and the nanocomposites developed via reduction aging, revealed considerable change in fracture mode with the appearance of the intragranular nano sized particles (on aging for 10 hrs or more). While Al_2O_3 exhibits the classical intergranular mode of fracture, the nanocomposites fracture via transgranular mode. Since fracture energy for cleavage (transgranular) fracture is comparatively higher than that for grain boundary (intergranular) fracture, this change in fracture mode may play some role in hindering crack propagation and concomitantly increasing the fracture toughness. The reasons behind such observations and their effects on the mechanical properties are currently under investigation.

4. CONCLUSION

The present work demonstrates how a novel process, based on solution treatment and aging in reducing atmosphere, can lead to the development of alumina based oxide-oxide micro/nano hybrid ceramic composite via cheaper conventional processing techniques. More specifically, the following conclusions can derived based on the current observations:

(a) Pressureless sintering at 1450°C for 5 hrs in air can lead to the development of solid solution of 10 wt.% Fe_2O_3 in Al_2O_3 (A10F).

(b) Aging in reducing atmosphere (N_2 - 4% H_2 gas mixture) at a temperature of 1450°C results in the precipitation of $FeAl_2O_4$ out of the solid solution.

(c) Aging treatments at 1450°C for 10 hrs lead to the formation of intragranular nano sized particles of $FeAl_2O_4$ in addition to micron sized particles along the matrix grain boundaries. However, considerable coarsening of the secondary phase particles ($FeAl_2O_4$) occurs on aging for elongated time period of ~ 50 hrs.

(d) The nano/micro composite, developed after aging for optimum duration (between 10 - 20 hrs at 1450°C), possess improved indentation toughness (by $\sim 45\%$) with respect to pure Al_2O_3.

REFERENCES
1. K. Niihara; *J. Ceram. Soc. Jap.* **The Centennial Memorial Issue 99[10]** (1991) 974-982.
2. I. P Shapiro, R. I. Todd, J. M. Titchmarsh, S. G. Roberts; *J. Eur. Ceram. Soc.* (article in press, 2009).
3. J. L. O. Merino, R. I. Todd, *Acta Mater.* **53[12]** (2005) 3345-3357.
4. A. Mukhopadhyay, B. Basu; *Intl. Mat. Rev.* **52[4]** (2007) 1-32.
5. K. Biswas, A. Mukhopadhyay, B. Basu, K. Chattopadhyay, *J. Mater. Res.* **22[6]** (2007) 1491-1501.
6. Y. Wang, T. Fujimoto, H. Maruyama, K. Koga; *J. Am. Ceram. Soc.* **83[4]** (2000) 933-36.
7. A. Mukhopadhyay; *Tribology - Materials, Surfaces & Interfaces* (article in press, 2009).
8. C. N. Walker, C. E. Borsa, R. I. Todd, R. W. Davidge, R. J. Brook; *Brit. Ceram. Proc.* **53** (1994) 249-264.
9. R. W. Davidge, P. C. Twiogg, F. L. Riley; *J. Eur. Ceram. Soc.* **16[7]** (1996) 799-802.
10. Y. Zhang, L. Wang, W. Jiang, L. Chen, G. Bai; *J. Eur. Ceram. Soc.* **26** (2006) 3393-3397.
11. H. Liu, C. Huang, J. Wang, X. Teng; *Mater. Res. Bull.* **41** (2006) 1215-1224.
12. S. E. Hsu, W. Kobes, M. E. Fine; *J. Am. Ceram. Soc.* **50[3]** (1967) 149-151.
13. R. A. Langensiepen, R. E. Tressler, P. R. Howell; *J. Mater. Sci.*, **18** (1983) 2771-76.
14. H. Y. Lee, Y. K. Paek, B. K. Lee, S. J. L. Kang; *J. Am. Ceram. Soc.* **78[8]** (1995) 2149-2152.
15. G. R. Anstis, P. Chantikul, B. R. Lawn, D. B. Marshall; *J. Am. Ceram. Soc.* **64** (1981) 553-557.
16. B. R. Lawn; Fracture of brittle solids. second edition; 1992, Cambridge University Press, Cambridge, UK.
17. M. S. Kaliszewski, G. Behrens, A. H. Heuer, M. C. Shaw, D. B. Marshall, G. W. Dransmann, R. W. Steinbrech; *J. Am. Ceram. Soc.* **77** (1994) 1185-1193.
18. K. Niihara, R. Morena, D. P. H Hasselman; *J Mater. Sci. Lett.* **1** (1982) 13-16.
19. S. Palmqvist; *Arch Eisenhuettenwes* **33** (1962) 629-333.
20. A. Mukhopadhyay, G. B. Raju, B. Basu, A. K. Suri; J. Eur. Ceram. Soc. **29** (2009) 505-516.
21. A. Mukhopadhyay, B. Basu, S. D. Bakshi, S. K. Mishra; *Intl. J. Ref. Met. Hard Mater.* **25** (2007) 179-188.
22. J. Tartaj, G. L. Messing; *J. Eur. Ceram. Soc.* **17** (1997) 719-725.

Table 1. Effect of aging duration on the microstructure development (distribution and size of secondary phase particles) and mechanical properties (Vickers hardness and Indentation toughness) of aged A10F (1450°C in reducing atmosphere).

Aging time (h)	Development of secondary phase particles (FeAl$_2$O$_4$)	Intergranular particles size (μm)	Intragranular particles size (nm)	Vickers Hardness (H$_{v2}$; GPa)	Indentation Toughness (K$_{1c}$; MPa m$^{1/2}$)
	Pure Al$_2$O$_3$ (sintered at 1450°C for 5 hrs in air)			20.2 ± 0.6	3.2 ± 0.3
0	Intergranular micron sized particles only	1.1 ± 0.6	-	19.3 ± 0.8	3.4 ± 0.3
5	Intergranular micron sized particles only	1.5 ± 0.8	-	19.1 ± 0.9	3.3 ± 0.8
10	Intergranular micron sized particles + Intragranular nano sized particles	1.7 ± 0.5	71 ± 27	19.9 ± 0.7	4.1 ± 0.4
15	Intergranular micron sized particles + Intragranular nano sized particles	1.9 ± 0.5	104 ± 38	19.2 ± 0.8	4.1 ± 0.3
20	Intergranular micron sized particles + Intragranular nano sized particles	2.2 ± 0.3	127 ± 34	19.2 ± 0.9	4.6 ± 0.2
50	Intergranular coarser micron sized particles + Intragranular finer sub micron sized particles	4.1 ± 0.8	321 ± 63	18.3 ± 1.5	3.5 ± 0.5

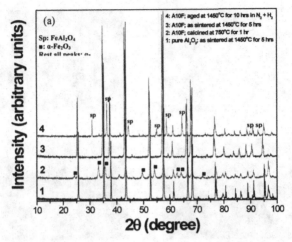

Fig. 1. (a) XRD patterns obtained from calcined A10F pellets and polished surfaces of as sintered pure Al_2O_3, solution treated A10F as well as A10F aged at 1450°C for 10 hrs in reducing atmosphere (N_2 + 4% H_2 gas mixture). Note the appearance of spinel phase on aging of the supersaturated A10F solid solution.

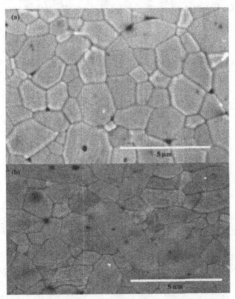

Fig. 2. Back scattered scanning electron micrographs obtained from polished and thermally etched surfaces of as sintered (a) pure Al_2O_3 and (b) A10F.

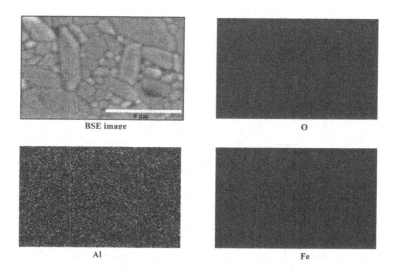

Fig. 3. X-ray elemental mapping, showing the uniform distribution of the various elements (Al, O, Fe) present on the polished thermally etched surface of as sintered A10F.

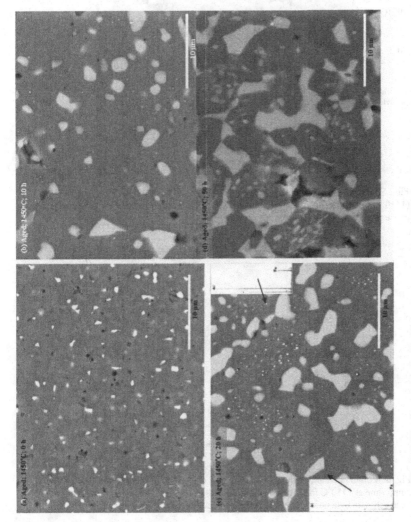

Fig. 4. BSE micrographs obtained from the as ground and polished surfaces (sample core) of A10F, aged in reducing atmosphere (N_2 + 4% H_2 gas mixture) at 1450°C for (a) 0 h; (b) 10 h; (c) 20 h; (d) 50 h. EDX spectra obtained from the matrix and secondary phase particles are presented in (c).

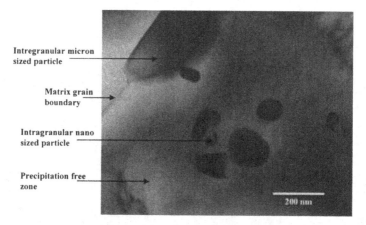

Fig. 5. Bright field transmission electron micrograph obtained from aged A10F (1450°C for 20 hrs in reducing atmosphere), showing the distribution of secondary phase particles (FeAl$_2$O$_4$).

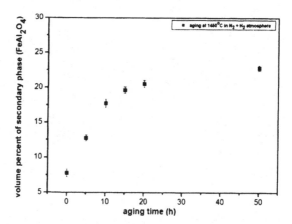

Fig. 6. Volume percent of the secondary phase (FeAl$_2$O$_4$) formed as a function of the aging time at 1450°C in reducing conditions (N$_2$ + 4% H$_2$). The volume fractions have been estimated via point counting method on back scattered SEM images.

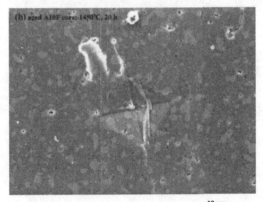

Fig. 7. Vickers indents, as obtained with a maximum load of 2 kg on the polished surfaces of (a) pure Al_2O_3 (b) A10F, aged at 1450°C for 20 hrs in reducing atmosphere. Note that radial cracking is suppressed for the nanocomposite (b) with respect to pure Al_2O_3 (a).

NANOSTRUCTURED ALUMINA COATINGS FORMED BY A DISSOLUTION/PRECIPITATION PROCESS USING AlN POWDER HYDROLYSIS

Andraz Kocjan*[1], Kristoffer Krnel[1], Peter Jevnikar[2], Tomaz Kosmac[1]
[1]Engineering Ceramics Department, Jozef Stefan Institute, Ljubljana, Slovenia
[2]Faculty of Medicine, University of Ljubljana, Ljubljana, Slovenia

ABSTRACT

The hydrolysis of aluminum nitride (AlN) powder was exploited in the preparation of nanostructured coatings on a sintered yttria-stabilized tetragonal zirconia (Y-TZP) ceramic substrate, with the aim to enhance the adhesion of dental cements to the ceramic core structure. The nanocrystalline coating formed using this method consists of γ-AlOOH (boehmite) in the form of 6-nm-thick and 240-nm-long interconnected lamellas. During a subsequent heat treatment in the temperature range from 600 °C to 1200 °C this coating was transformed to various transient aluminas without any noticeable change in the morphology, but its bonding to the substrate was significantly improved. By varying the precipitating conditions, such as the temperature and the time, it was also possible to synthesize coatings differing in composition and morphology, resulting in nano- or micron-sized particles. This nanosized alumina coating has the potential to improve the adhesion of composite dental cements to the sintered Y-TZP ceramic.

INTRODUCTION

The reaction of AlN powder with water has been known for a long time. In the presence of water, AlN will decompose, forming aluminum hydroxide and ammonia. The hydrolysis of AlN powder was mainly studied in order to prevent the reaction, since AlN ceramics are used as a substrate material for power circuits and as a packaging material for integrated circuits.[1] Only a few authors have investigated the mechanisms of aluminum hydroxides formation during the reaction of AlN powder with water.[2-5] Bowen et al.[2] investigated the hydrolysis at room temperature and proposed the following reaction scheme:

$$AlN + 2H_2O \rightarrow AlOOH_{amorph} + NH_3 \qquad (1)$$
$$NH_3 + H_2O \rightarrow NH_4^+OH^- \qquad (2)$$
$$AlOOH_{amorph} + H_2O \rightarrow Al(OH)_3 \qquad (3)$$

The AlN powder first reacts with water to form amorphous aluminum oxyhydroxide (pseudoboehmite phase, γ-AlOOH), which later re-crystallizes as aluminum 3-hydroxide (bayerite, α-Al(OH)₃) by the dissolution/precipitation process, according to reaction (3).[6,7] The kinetics of AlN hydrolysis was described using an unreacted-core model, proposed by Levenspiel,[8] and the chemical reaction at the product-layer/unreacted-core interface was proposed to be the rate-controlling step during the initial stage of the reaction.[2,5]

At elevated temperatures, the starting temperature and especially the ageing time in the mother liquor, i.e., the time of the hydrolysis, strongly influence the reaction products and their morphology.[3,4,5,9]

Svedberg et al.[3] studied the corrosion of AlN in aqueous solutions at various constant pH values (5, 8, 11, 14) heated to 85 °C for 1 h. In all pH regimes the pseudoboehmite and bayerite/gibbsite phases, in various proportions, were detected with XRD.

Fukumoto et al.[4] investigated the hydrolysis behavior of AlN powder in water at room temperature and elevated temperatures up to 100 °C. According to these authors the hydrolysis behavior changes at 78 °C: below this temperature, crystalline bayerite is the predominant phase, while above 78 °C, crystalline boehmite is formed.

In our recent work we also found that, at temperatures higher than room temperature, the first crystalline product is boehmite, and with a prolonged ageing time the bayerite conversion takes place by the dissolution of pseudoboehmite and the recrystallization of bayerite.[9]

More recently, the AlN powder hydrolysis at elevated temperatures has been exploited in the Hydrolysis Assisted Solidification (HAS) forming process, in which the hydrolysis reactions provoke the solidification of the aqueous ceramic suspension in an impermeable mould. In this process a few percent of AlN powder is added to the aqueous slurry of the host ceramic powder. After homogenization, the suspension is cast into a closed, nonporous mould, where the hydrolysis is thermally activated. Due to the water consumption, the formation of aluminum hydroxides on the surface of the parent ceramic particles and the change in the ionic strength of the suspension during the hydrolysis of AlN powder, the viscosity of the host slurry is increased to such an extent that a saturated solid body is formed.[10-12] The precipitation of the aluminum hydroxides on the surface of the host ceramic particles in the ceramic suspension during the solidification in the HAS process also indicated that the formation of nanostructured aluminum hydroxide coatings on hydrophilic surfaces with the use of AlN powder hydrolysis is a feasible process.

In this work the preparation of nanostructured alumina coatings by exploiting the AlN powder hydrolysis is presented. The influence of precipitation conditions, such as the time and temperature, on the morphology of the precipitated boehmite coatings was studied and the crystal structure of the coatings before and after the heat treatment was analyzed. In addition, the potential of this coating to improve the adhesion of dental cements to the sintered Y-TZP ceramic, depending on the surface treatment, was investigated. This is because Y-TZP is widely used as a core material in restorative dentistry, due to its high strength, biocompatibility and favorable optical properties. However, the adhesion of the dental cement to the Y-TZP surface is poor.[13] For this reason sandblasting is commercially used to increase the surface roughness of the ceramics and thereafter improve the adhesion of the dental cement.[14,15] As an alternative, chemical modification of the zirconia surface, i.e., silanization and others, is being investigated to improve the binding between the dental cement and the zirconia surface.[16,17]

EXPERIMENTAL

The AlN powder used in the experimental work was AlN Grade C powder (H.C. Starck, Berlin, Germany) with a nominal particle size of 1.2 μm, an oxygen content of 2.2 wt.%, and a specific surface area of 3.2 m^2/g. The hydrolysis tests were carried out in dilute suspensions containing 3 wt% of AlN in deionized water. In these tests the water was preheated with an electric heater under constant stirring to the desired temperature (40 °C, 50 °C, 60 °C, 70 °C, 80 °C and 90 °C) and then the AlN powder was added to the water. The pH and the temperature were measured versus time using a combined glass-electrode/Pt 1000 pH meter (Metrohm 827) equipped with an accurate thermoelement.

The crystalline phases of the hydrolysis products were characterized using X-ray diffraction (XRD), scanning electron microscopy (SEM) and transmission electron microscopy (TEM).

The zirconia substrates were produced from commercially available ready-to-press TZ-3YSB-E (Tosoh Corp, Tokyo, Japan) powder containing 3 mol.% yttria and 0.25 wt.% alumina in the solid solution to hinder the ageing process. The discs were fabricated by uni-axial dry pressing, and sintering at 1500 °C for 2 hours in air. One side of each disc was ground and polished with 6-μm and 3-μm diamond pastes using a standard metallographic procedure.

In the preparation of aluminate coatings, the zirconia disc was inserted into preheated deionized water and the AlN powder was added under constant stirring and heating. The immersion time ranged from a few minutes to several hours. Afterwards, the substrate was removed from the slurry, dried in air at 100 °C for 30 minutes and subjected to a heat treatment in dry air for 1 hour at 900 °C, 1200 °C and 1300 °C. The heating rate was 10 °C/min.

For the characterization of the coated Y-TZP substrates SEM and TEM analyses were employed.

The adhesion of the coating to the Y-TZP substrate before and after the heat treatment was evaluated using a simple "Scotch tape" test. In addition, the shear bond strength of a commonly used composite dental cement to uncoated and coated Y-TZP substrates, differing in the surface pre-treatment, was determined according to the ISO 6782 (2008) standard.

RESULTS and DISCUSSION

The hydrolysis of AlN powder

The variation of pH versus time of the AlN-powder water suspension for various starting temperatures is presented in Figure 1. The starting pH of the deionized water ranged from 4 to 7, because of the adsorption of CO_2 from the air. When the AlN powder was dispersed in water preheated to 40 °C, 50 °C and 70 °C, the starting pH value initially decreased even more (revealed by the inset graph in Figure 1.), presumably due to the uptake of additional CO_2 during the vigorous stirring needed to homogenize the suspension. The higher starting pH values for the higher starting temperatures correspond to the solubility of the CO_2 in water, i.e., the solubility of the CO_2 in the water is decreasing with the increasing water temperature. After the addition of the AlN powder to the preheated water the pH started to increase and reached an equilibrium pH value within a few minutes to several hours, depending on the starting temperature. A higher starting temperature evidently speeds up the reaction of the AlN powder with water, also lowering the equilibrium pH value, presumably due to the decreasing solubility of ammonia with the increasing temperature.[9]

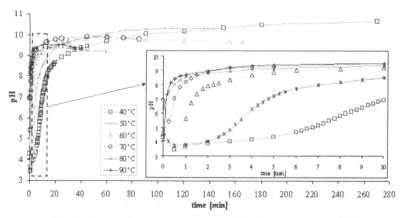

Figure 1: pH versus time for the hydrolysis of a 3 wt.% AlN suspension in deionized water at various starting temperatures.

Since the reaction of AlN powder with water is an exothermic process,[18] we also monitored the temperature increase versus time for the AlN powder suspension for various starting temperatures. The results are presented in Figure 2. For the starting hydrolysis temperatures of 50 °C, 60 °C and 70 °C, the temperature increase (ΔT), i.e., the difference between the highest measured temperature and the starting temperature, is about 18 °C. The ΔT value for the AlN powder hydrolyzed at 80 °C is 15 °C, and at 90 °C it is only 8 °C, because the temperature reached the boiling point of the suspension. The time needed to reach the maximum ΔT value (ΔT_{MAX}; the black lines perpendicular to the ΔT curve) for the starting temperatures of 50 °C, 60 °C and 70 °C was 49 min, 27 min and 13 min, respectively. For the 80 °C and 90 °C curves, the ΔT_{MAX} (the dashed black lines perpendicular to the ΔT curve) values were estimated to be 9 min and 7 min, respectively. After reaching ΔT_{MAX}, the

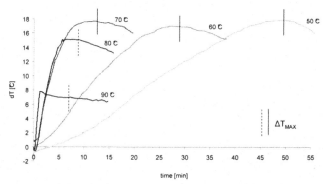

Figure 2: dT versus time for the hydrolysis of a 3 wt% AlN powder suspension in deionized water at 50 °C, 60 °C, 70 °C, 80 °C and 90 °C.

slurries were filtered, washed with isopropanol and dried for subsequent XRD analyses. In all cases only boehmite and unreacted AlN were indentified from the corresponding XRD pattern. With even longer ageing times, peaks of bayerite were also detected in the XRD pattern, in agreement with previously reported studies.[3,4,5,9] A typical SEM micrograph of AlN powder hydrolyzed at 50 °C for 2 hours is presented in Figure 3, showing small, agglomerated, nanostructured boehmite (AlOOH) lamellas and large, elongated bayerite (Al(OH)$_3$) crystals. Bayerite particles usually occur as somatoids, which are defined as bodies of uniform shape that are not enclosed by crystal faces. These shapes resemble hour glasses, cones, or spindles.[6] On the other hand, the crystallites of lamellar boehmite have at least two growth dimensions, ranging from 5 nm to 50 nm. The thickness of the crystallites constitutes the third dimension, generally ranging from about 2 nm to 10 nm.[19]

From the above results it can be assumed that at ΔT_{MAX} the extensive boehmite formation is terminated, and with prolonged hydrolysis times bayerite will be formed at a slower reaction rate. The vigorous reaction during the initial stages of hydrolysis, i.e., prior to the ΔT_{MAX} being reached, should be accompanied by a rapid increase in the concentration of dissolved aluminum (poly)cations in the suspension, which favor the nucleation of boehmite.[20,21] With an appropriate amount of AlN powder in the suspension it is likely that the degree of supersaturation will be sufficiently high to provoke a heterogeneous nucleation on the surface of an immersed hydrophilic substrate. This assumption was experimentally confirmed by monitoring the boehmite formation on a polished Y-TZP substrate immersed in the 3 wt.% AlN powder suspension at 70 °C for various periods of time.

Figure 3: SEM micrograph of AlN powder after hydrolysis for 2 hours at 50 °C, showing large, elongated bayerite (Al(OH)$_3$) crystals and small, agglomerated, nanostructured boehmite (AlOOH) lamellas.

Coating morphologies

The SEM micrographs in Figure 4 reveal that the nucleation of the boehmite lamellas after soaking the substrate for 1 minute (Figure 4a) had already begun. After 3 minutes, the lamellas became interlocked (Figure 4b), and eventually, after 15 min of immersion, the

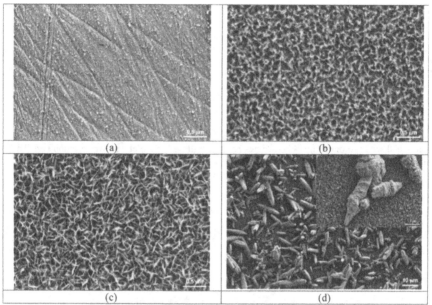

Figure 4: SEM micrographs of the precipitated boehmite and bayerite coatings on the polished Y-TZP surface using AlN powder hydrolysis at 70 °C for (a) 1 min, (b) 3 min, (c) 15 min and (d) 4 h.

perpendicular growth of the nanostructured boehmite lamellas was complete (Figure 4c), as predicted from the ΔT versus time diagram (Figure 3). With longer reaction times, large, elongated, micron-sized bayerite particles started to grow on the coated Y-TZP substrate, whereas the boehmite coating remained unchanged (Figure 4d).

The boehmite coatings prepared at other starting temperatures and the corresponding ΔT_{MAX} times, as estimated from Figure 3, are shown in Figure 5. The micrographs indicate that the thickness of the lamellas, which was relatively insensitive to the starting temperature, was around 6 nm, which was later confirmed by the TEM analysis. The thickness of the boehmite coatings, as estimated from the SEM micrographs of tilted cross-sections, was about 240 nm, and was also not influenced by the starting temperature. In contrast, the surface density of the lamellas, i.e., the number of lamellas per unit area, was found to be temperature dependent and decreased with higher hydrolysis temperatures. At the same time, their width and distinctness were increased with higher temperatures.

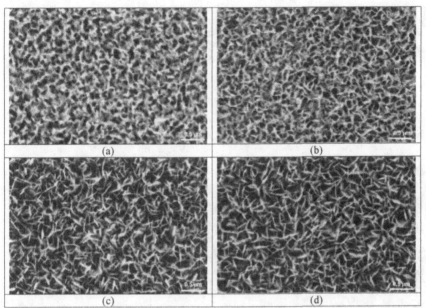

Figure 5: SEM micrographs of the precipitated boehmite coatings on the Y-TZP surface using AlN powder hydrolysis at (a) 50 °C for 50 min, (b) 60 °C for 30 min, (c) 80 °C for 10 min and (d) 90 °C for 7 min.

The effects of heat treatment

After drying, the adhesion of the boehmite coating to the Y-TZP surface was poor and it could be easily peeled off during the "Scotch tape" test. In order to improve the adhesion, a heat treatment of the coated substrates was performed, during which the topotactic transformation of the boehmite to various transient aluminas occurred. SEM micrographs and the corresponding results of the TEM analyses of the boehmite precipitates followed by a heat treatment are shown in Figure 6. The morphology of the coating remained basically unchanged up to 900 °C (Figure 6b), whereas at higher temperatures the lamellas started to coarsen and sinter (Figure 6c), until at 1300 °C (Figure 6d) they practically disappeared, forming a thin discontinuous film. The structure of the as-precipitated polycrystalline lamellas corresponds to boehmite (orthorhombic, γ-AlOOH, Figure 6a). The corresponding TEM analysis of the boehmite precipitates after heat treatment revealed that at 900 °C delta alumina (tetragonal, δ-Al$_2$O$_3$) is formed, at 1000 °C to 1200 °C it transforms to theta alumina (monoclinic, θ-Al$_2$O$_3$), whereas at 1300 °C alpha alumina thin film (rhomboedric, α-Al$_2$O$_3$) is formed.

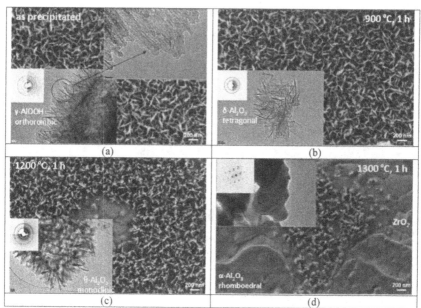

Figure 6: SEM micrographs of the (a) as-precipitated boehmite coating using 3 % AlN powder hydrolysis at 90 °C for 7 min, (b) coating after heat treatment at 900 °C for 1 h, (c) coating after heat treatment at 1200 °C for 1 h and (d) coating after heat treatment at 1300 °C for 1 h on the Y-TZP surface. The inset pictures represent the corresponding TEM and SAED analysis of the precipitates.

Bond strengths with dental cement

In order to explore the potential of this coating for establishing a stronger bond between the ceramic core material and the luting agent in restorative dentistry, the adhesion of dental cements to the Y-TZP ceramics was studied using the shear-bond test, following the ISO 6872 (2008) standard. The adhesion of a modern composite cement with an active phosphate monomer (MDP) on the differently prepared surfaces, i.e., as-sintered, polished and sandblasted, of Y-TZP ceramic was investigated and the results obtained with commercial MDP cement (Rely X Unicem, 3M ESPE, USA) are presented. As shown in Fig. 7, this coating indeed has the potential to substantially enhance the adhesive bond by a factor of 2–4, depending on the mechanical pre-treatment of the substrate.

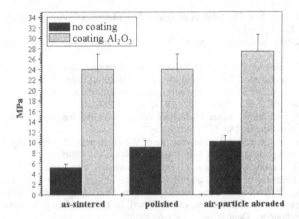

Figure 7: Shear bond strength of the composite dental cement Rely X Unicem to the surface of the Y-TZP.

CONCLUSIONS

The hydrolysis of AlN powder dispersed in deionized water at elevated temperatures is accompanied by a rapid increase in pH and temperature. A vigorous reaction during the initial stages of hydrolysis, i.e., prior the ΔT_{MAX} value, provokes the heterogeneous nucleation of boehmite onto the surface of the immersed Y-TZP substrate. This can be exploited for the formation of nanostructured boehmite coatings. Irrespective of the starting temperature, the coating consisted of 6-nm-thick and 240-nm-long interconnected polycrystalline γ-AlOOH lamellas. The surface density and the width of these lamellas depend on the initial temperature of the aqueous AlN slurry. During a subsequent heat treatment up to 900 °C, these coatings are transformed to a transient alumina without any noticeable change in the morphology, but their bonding to the substrate after the heat treatment is improved. This nanosized alumina coating has the potential to significantly improve the adhesion of composite dental cements to the sintered Y-TZP ceramic.

REFERENCES

[1]L. M. Sheppard, Aluminum nitride: a versatile but challenging material. *Am. Ceram. Soc. Bull.*, **69**, 1801–1812 (1990).
[2]P. Bowen, J.G. Highfield, A. Mocellin, and T.A. Ring, Degradation of Aluminum Nitride Powder in an Aqueous Environment, *J. Am. Ceram. Soc.*, **73** [3], 724-728 (1990).
[3]L. M. Svedberg, K. C. Arndt, and M. J. Cima, Corrosion of aluminum nitride (AlN) in aqueous cleaning solutions, *J. Am. Ceram. Soc.*, **83**, 41–46 (2000).
[4]S. Fukumoto, T. Hookabe, and H. Tsubakino, Hydrolysis behavior of aluminum nitride in various solutions, *J. Mater. Sci.*, **35**, 2743–2748 (2000).
[5]K. Krnel, G. Drazic, and T. Kosmac, Degradation of AlN powder in aqueous environments, *J. Mater. Res.*, **19**, 1157–1163 (2004).
[6]K. Wefers, and C. Misra, *Oxides and Hydroxides of Aluminum*, Technical Paper No. 19 (revised 1987) available from Alcoa, Pittsburg, PA.

[7]T. Graziani, and A. Belosi, Degradation of Dense AlN Materials in Aqueous Environments, *Mater. Chem. Phys.*, **35**, 43-48 (1993).

[8]O. Levenspiel, *Chemical Reaction*, Engineering (2nd ed.). John Wiley & Sons, New York, pp. 357–408 (1972).

[9]A. Kocjan, K. Krnel, and T. Kosmac, The influence of temperature and time on the AlN powder hydrolysis reaction products, *J. Eur. Ceram. Soc.*, **28**, 1003-1008 (2008).

[10]T. Kosmac, S. Novak, and M. Sajko, Hydrolysis-assisted solidification (HAS): a new setting concept for ceramic net-shaping, *J. Eur. Ceram. Soc.*, **17**, 427–432 (1997).

[11]T. Kosmac, S. Novak, and K. Krnel, Hydrolysis assisted solidification process and its use in ceramic wet forming, *Z. Metallkd.*, **92**, 150–157 (2001).

[12]T. Kosmac, The densification and microstructure of Y-TZP ceramics formed using the hydrolysis-assisted solidification process, *J. Am. Ceram. Soc.*, **88**, 1444–1447 (2005).

[13]C. Piconi, and G. Maccauro, Zirconia as a ceramic biomaterial, *Biomaterials*, **20**, 1-25 (1999).

[14]M. Wolfart, F. Lehman, S. Wolfart, and M. Kern Durability of the resin bond strength to zirconia ceramic after using different surface conditioning methods. *Dent Mater* **23**, 45-50 (2007).

[15]M. Uo, G. Sjogren, A. Sundh, M. Goto, ynd F. Watari, Effect of surface condition of dental zirconia ceramic Denzir on bonding. *Dent Mater J*, **25**, 1-6 (2006);

[16]M. Blatz, G. Chiche, S. Holst, and A. Sadan, Influence of surface treatment and simulated aging on bond strengths of luting agents to zirconia. *Quint Int*, **38**, 745-53 (2007).

[17]J. P. Matinlinna, T. Heikkinen, M. Ozcan, L. V. J. Lassila, and P. K. Vallittu, Evaluation of resin adhesion to zirconia ceramic using some organosilanes, *Dental Mat*, **22**, 824-831 (2006)

[18]W.M. Mobley, *Colloidal Properties, Processing and Characterization of Aluminum Nitride Suspensions*, Ph.D. Thesis, Alfred University, Alfred, New York (1996), p. 110.

[19]P. Euzen, P. Raybaud, X. Krokidis, H. Toulhoat, J. L. Le Loarer, J. P. Jolivet, Froidefond, "Alumina"; pp. 1591–677 in Hand Book of Porous Solids, Edited by F. Schuth, K. S. W. Sing, and J. Weitkamp. Wiley, Chichester, 2002.

[20]J. P. Jolivet, *Metal Oxide Chemistry and Synthesis—From Solution to Solid State*, Wiley, Chichester, (2000).

[21]M. Henry, J. P. Jolivet, and J. Livage, *Aqueous Chemistry of Metal Cations: Hydrolysis, Condensation, and Complexation*, Struct. Bonding, **77**, 153–206 (1992).

SYNTHESIS OF ALUMINUM NITRIDE NANOSIZED POWDER BY PULSED WIRE DISCHARGE WITHOUT AMMONIA

Yoshinori Tokoi[1], Tsuneo Suzuki[2], Tadachika Nakayama[2], Hisayuki Suematsu[2], Futao Kaneko[1] and Koichi Niihara[2]

[1]Department of Electrical and Electronic Engineering, Niigata University, 2-8050 Ikarashi Nishiku, Niigata, 950-2181, Japan

[2]Extreme Energy-Density Research Institute, Nagaoka University of Technology, 1603-1 Kamitomioka-cho, Nagaoka 940-2188, Japan

ABSTRACT

High-purity aluminum nitride (AlN) nanosized powders synthesized by pulsed wire discharge (PWD) using aluminum wires in nitrogen gas and nitrogen/argon mixed gas. The synthesis was carried out at various relative energies (K) of 2.7-289, ambient gas pressures (P) of 10-100 kPa and nitrogen partial pressure of nitrogen/argon mixed gas (P_{N2}) of 0-10kPa, where K was the ratio of the charged energy of the capacitor to the vaporization energy in the wire. The AlN content (C_{AlN}) in the synthesized powders were calculated from X-ray diffraction analysis and the synthesized powders were observed by a transmission electron microscopy. The increase in C_{AlN} resulted from the increase in K, the decrease in P, and the increase in P_{N2}. The highest C_{AlN} of 97wt% with a median particle diameter of 6nm was obtained at K=289, P=10kPa and P_{N2}=10kPa. In the present study, it was clarified that high K and low P were necessary for the synthesis of high-purity AlN nanosized powder by PWD in nitrogen gas.

INTRODUCTION

Aluminum nitride (AlN) is a very important compound for applications in integrated circuits, power modules, microelectronics and light emitting devices, due to its high thermal conductivity, low dielectric constant, low thermal expansion coefficient, good mechanical strength as well as chemical and thermal stability, and wide direct band gap[1,2]. AlN nanosized powders are also expected to be used as raw materials for low temperature sintering of the electrical modules and light emitting applications. For these applications, high-purity, high-crystallinity and small particle size are required[3].

High-purity and high-crystallinity AlN nanosized powder can be synthesized through rapid quenching of materials from the gas phase by several techniques such as pulsed laser ablation (PLA), ion beam evaporation (IBE), DC arc discharge, RF plasma and transferred plasma[4-8]. Recently, environmentally-friendly synthesis methods are required worldwide. However, most of these methods are using ammonia (NH₃) to synthesize high-purity AlN nanosized powders. Therefore, this research is attempted to develop a novel synthesis method of AlN nanosized powder without NH₃, because NH₃ requires careful handling because of its flammable, corrosive and toxic properties.

Pulsed wire discharge (PWD) is one of the physical vapor methods for the preparation of nanosized powders using plasma/vapor and one of the applications of pulsed power technology[9-13]. The plasma/vapor is generated by the rapid heating of a thin metal wire, due to the high-pulsed current which is produced by an electrical discharge circuit. During the rapid cooling of the plasma/vapor, many nanosized powders are formed by grain growth and chemical reaction after nucleation, which result from collision between the plasma/vapor and ambient gas molecules. This method has significant advantages, including high-energy conversion efficiency and coproduct-free synthesis of nanosized powders[14]. The particle size is controlled by the pressure of ambient gas (P) and the relative energy (K)[13,15,16], where K is the ratio of the charged energy of the capacitor in the electrical discharge circuit to the vaporization energy in the wire which is calculated from the standard formation enthalpy of a gas phase and the volume of the wire.

In previous research of the AlN nanosized powder synthesized by PWD, high-purity AlN nanosized powders were synthesized using aluminum (Al) wire in N_2/NH_3 mixed gas. The AlN content (C_{AlN}) was increased with increasing P, K and NH_3 concentration of N_2/NH_3 mixed gas[13,14,16-18]. The highest AlN content (C_{AlN}) of 98wt% with an average particle size of 30nm was obtained at an NH_3 concentration of 20% and $K=3.8$[18].

On the other hand, AlN nanosized powders synthesized in N_2 gas at $P=100$kPa and $K=2$, and the C_{AlN} in the prepared powder was less than 20wt%[14,17]. However, Kotov et al. succeeded in the synthesis of AlN nanosized powder in N_2 gas, and reported that the C_{AlN} was increased with increasing P and K[15,19]. The highest C_{AlN} of 86wt% was obtained at very high-pressure of $P=300$kPa and $K=1.75$[19]. In this condition, small particles were not synthesized because the particle size is increased with increasing P[13]. Therefore, the indicator of the small particle size and high-purity AlN nanosized powder synthesized in N_2 was not obtained. Additionally, the nitriding process during the formation of AlN nanosized powder has not yet been clarified.

In this article, we report experimental results of the synthesis of AlN nanosized powder by PWD using Al wire in N_2 and/or N_2/Ar mixed gas at various K, P and N_2 partial pressure in N_2/Ar mixed gas (P_{N2}), and the relationship of C_{AlN}, K P and P_{N2}. Using these results, we discuss the influence of K, P and P_{N2} on the C_{AlN} in PWD and the nitriding process during the AlN nanosized powder synthesized by PWD in N_2 gas.

EXPERIMENTAL PROCEDURE

Figure1 shows a schematic illustration of the experimental setup used for synthesis AlN nanosized powders. The electrical discharge circuit for generated high-pulsed current was composed of a capacitor bank, a high-voltage DC power supply (H. V.) and a spark gap. The experimental procedure for AlN nanosized powder synthesis is as follows. A chamber was evacuated using a vacuum pump, and was then filled with N_2 gas and/or N_2/Ar mixed gas, after an Al wire was placed between the electrodes in the chamber and a membrane filter was installed between the chamber and a vacuum pump. After the capacitor bank was charged by the high-voltage DC power supply, the Al wire evaporated and ionized due to the high-pulsed current by closing the spark gap. The powders floating in the chamber were collected on the surface of a membrane filter with a pore size of 0.1 μm by evacuating the gas through the membrane filter using the vacuum pump.

In the experiment, the voltage ($V(t)$) between the electrodes and the current ($i(t)$) of the circuit during PWD were measured using a digitized oscilloscope with two high-voltage probes (V_L and V_H) and a current transformer (C.T.), respectively. The energy deposition ($E(t)$) during PWD was calculated from the measured $V(t)$ and $i(t)$[20], where $V(t)$ was obtained from the difference between the two probe signals of $V_H(t)$ and $V_L(t)$.

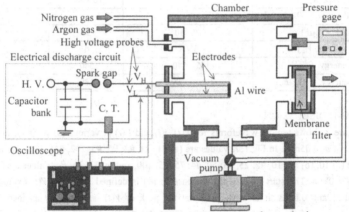

Figure 1. Experimental setup for AlN nanosized powder synthesis.

The synthesized nanosized powders were analyzed using a powder X-ray diffractometer (XRD; Rigaku RINT2000, Cu-Kα, 50 kV, 300 mA) and a transmission electron microscope (TEM; Jeol JEM-2000FM). The phases in the synthesized nanosized powders were identified from the peak positions in the XRD patterns. In addition, the C_{AlN} was obtained by the area ratio of the AlN (100) peak and the Al (111) peak after peak separation[14]. The median particle diameter (D_{50}) and the geometric standard deviation (σ_g) were calculated from the particle size distribution which was evaluated by measuring the diameter of more than 1000 particles from bright-field TEM images[21].

The experimental conditions for AlN nanosized powder synthesized at various (a) K with constant P in N_2, (b) P with constant K in N_2 and (c) P_{N2} with constant K and P are summarized in Table I. The charged energy of the capacitor bank (E_c) was calculated by the capacitance (C) of the capacitor bank and the charged voltage (V_c). The vaporization energy of the Al wire (E_v) was changed by the Al wire diameter of 0.3-0.05 mm[22]. K was controlled by E_c and E_v, and it was changed from 2.7 to 289. In the condition of (c), the total pressure of N_2/Ar mixed gas was set constant at 10 kPa, and the

N_2 partial pressure of the N_2/Ar mixed gas was changed from 0 to 10kPa.

Table I. Experimental conditions for AIN nanosized powders synthesis

	Species	Aluminum					
Wire [mm]	Length, l [mm]	25					
		(a)				**(b)**	**(c)**
	Diameter, d [mm]	0.3	0.2	0.1	0.05	0.05	0.05
Vaporization energy, E_v [J]		67.2	29.9	7.5	1.9	1.9	1.9
Gas species		N_2				N_2	N_2/Ar mixed
Pressure, P [kPa]		10				10 - 100	10
N_2 partial pressure, P_{N2} [kPa]		-				-	0 - 10
Capacitance, C [μF]		10 - 30				30	30
Charged voltage, V_c [kV]		3 - 6				6	6
Charged energy, E_c [J]		45 - 540				540	540
Relative energy, K (E_c/E_v)		2.7 - 289				289	289

RESULTS AND DISCUSSION

Figure 2 shows typical waveforms for $V(t)$, $i(t)$ and $E(t)$ at various (a) K, (b) P and (c) P_{N2} during PWD of d=0.05mm. In Fig.1, K values are (a) #1: 24, #2: 96, #3: 192, #4: 289, (b) and (c) 289. P values are (a) 10, (b) #1: 10, #2: 25, #3: 50, #4: 75, #5: 100, and (c) 10kPa. P_{N2} values are (c) #1: 10, #2: 5 and #3: 0kPa. The start time of the increase in $i(t)$ is defined as t=0μs. $V(t)$ exhibits a peak between t=0.25 and 0.81μs. It is considered that the peak of $V(t)$ is caused by an increase in the electrical resistance of the Al wire between the electrodes due to Al wire vaporization with ohmic heating, and by a decrease in the electrical resistance between electrodes due to the formation of the Al plasma[20]. After the formation of the plasma at t=0.25-0.81μs, the damped oscillation is observed. During damped oscillation, an arc discharge is generated between the electrodes through the Al and/or nitrogen/argon plasma. Therefore, the heat of materials in PWD consists of two stages: (1) heating of solid/liquid/vapor state of the wire (2) heating of plasma included wire materials and ambient gas. In the case of K change, the time of the voltage peak becomes earlier with increasing K, because the inclination of the increase in $i(t)$ becomes steeper with increasing K and/or E_c. In the case of P change, $E(t)$ is approximately the same value, because $V(t)$ and $i(t)$ are same waveforms. In the case of P_{N2} change, the decrease in P_{N2} results in the early occurrence in the time of the voltage peak from t=0.37 to 0.25μs and the suppressed height of the voltage peak from 10 to 5.6kV. It is considered that the above result is caused by the decrease in the breakdown voltage between electrodes with decreasing P_{N2} because the electrical breakdown voltage of Ar gas is lower than that in N_2 gas[23]. In Fig.2, we define the deposited energy in the wire (E_w) and the total deposited energy (E_t) up to the time of the voltage peak around t=0.25-0.81μs and up to the total time of t=80μs, respectively[20].

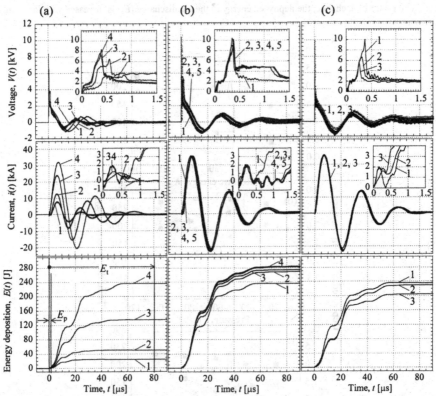

Figure 2. $V(t)$, $i(t)$ and $E(t)$ during PWD of d=0.05mm at various (a) K, (b) P and (c) P_{N2}. K values are (a) #1: 24, #2: 96, #3: 192, #4: 289, (b) and (c) 289. P values are (a) 10, (b) #1: 10, #2: 25, #3: 50, #4: 75, #5: 100, and (c) 10kPa. P_{N2} values are (c) #1: 10, #2: 5 and #3: 0kPa. E_w and E_t are deposited energy in the wire and total deposited energy, respectively.

Figure 3 shows the dependence of E_w and E_t and E_a on (a) K, (b) P and (c) P_{N2} with d=0.05mm. With increased K, E_w becomes saturated and E_t increases linearly. Additionally, E_w exceeds E_v at more than K=100. From this result, it is consider that the used Al wire of d=0.05mm is evaporated completely at more than K=100. In the case of P change, E_w and E_t are almost constant and E_w exceeds E_v at all values of P. On the other hand, in the case of P_{N2} change, E_w and E_t are decreased with decreasing P_{N2}. The lowest E_w of 0.74J is obtained at P_{N2}=0kPa and this value is less than half of E_v. In this case, it is considered that the wire was not completely evaporated whole of the wire. The majority of E_t is consumed during arc discharge after the wire heating, because E_w is very small compared with

E_t. In the case of K change, the deposited energy in the arc discharge (E_a) is increased with increasing K.

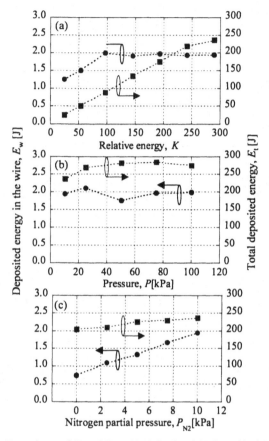

Figure 3. Dependence of E_w and E_t on (a) K, (b) P and (c) P_{N2} with d=0.05mm.

Figure 4 shows XRD patterns of the nanosized powders synthesized at various (a) K with P=10kPa in N_2 gas, (b) P with K=289 in N_2 gas and (c) P_{N2} with K=289 and P=10kPa. The lower vertical line patterns in the figure show the diffraction peak positions and the relative intensities for Al and AlN according to the International Center for Diffraction Data (ICDD)[24,25]. The peaks of the XRD patterns matched those of the ICDD for Al and AlN, and the synthesized nanosized powders include AlN and Al.

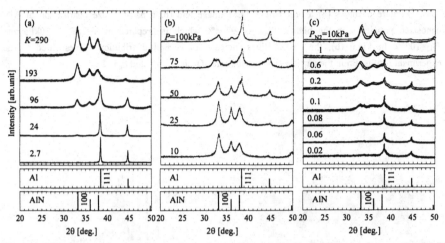

Figure 4. XRD patterns for particles synthesized at various (a) K with P=10kPa in nitrogen gas, (b) P with K=289 in nitrogen gas and (c) P_{N2} with K=289 and P=10kPa.

Figure 5 shows the dependence of the C_{AIN} on (a) K, (b) P and (c) P_{N2}. In Fig.5 (a), the C_{AIN} is increased from 2 to 97wt% with increasing K from 2.7 to 289. In the case of P change, the C_{AIN} is decreased from 97 to 26wt% with increasing P from 10 to 100kPa. In the case of P_{N2} change, the C_{AIN} is decreased with decreasing P_{N2} and it displays two inclinations: (1) steep inclination from P_{N2}=0 to 20% and (2) gradual inclination from P_{N2}=2 to 10kPa. The C_{AIN} of 0wt% is obtained at less than P_{N2}=0.04kPa.

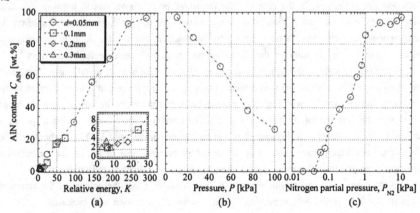

Figure 5. Dependence of the C_{AIN} on (a) K at d=0.05-0.3mm, (b) P and (c) P_{N2}.

Figure 6 shows (a) a bright-field TEM image and (b) a particle size distribution of AlN nanosized powder with the highest C_{AlN} of 97wt% which is prepared at $K=289$, $P=10$kPa and $P_{N2}=100\%$. In Fig.6 (b), the dashed lines indicate log-normal distribution curves fitted to the experiment results. From the particle size distributions, D_{50} is 6.3 nm and σ_g is 1.6.

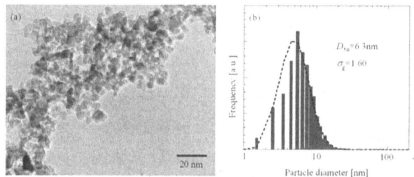

Figure 6. (a) Bright-field TEM image and (b) particle size distribution of AlN nanosized powder with highest C_{AlN} of 97wt% which is prepared at $K=289$ and $P=10$kPa in nitrogen gas. D_{50} and σ_g are median particle diameter and geometric standard deviation, respectively.

Possible reasons for the increase in the C_{AlN} with increasing K and decreasing P are discussed. In PWD, grain growth and chemical reaction occur after nucleation simultaneously during the rapid cooling of the plasma/vapor, which result from collision between the plasma/vapor of the wire material and ambient gas molecules. In the case of AlN nanosized powder synthesis in nitrogen gas, it is considered that the amount and lifetime of the activated nitrogen species during the cooling of the Al plasma/vapor is very important factor for the nitriding reaction, because the dissociation energy of nitrogen is higher that in ammonia and the lifetime of the activated nitrogen species is very short[26].

In the previous study, during the arc discharge after the wire heating, metal ions and activated chemical species of the ambient gas were observed by a spectroscopic method[27]. In the present study, the activated nitrogen species was generated around the Al plasma/vapor, because the arc discharge occurred after the wire heating as shown in Fig. (2). The amount of the activated nitrogen species is influenced by the parameter of K, because E_a is changed by the parameter of K. On the other hand, the lifetime of the activated nitrogen species is influenced by the parameter of P, because the collision frequency of nitrogen molecules and activated nitrogen species is changed by the parameter of P.

In the case of K change with constant P, it is considered that the amount of the activated nitrogen species is increased with increasing E_a which is lineally increased with increasing K. On the other hand, lifetime of the activated nitrogen species is almost constant, because P is constant at 10kPa.

Therefore, it is considered that the cause of the increase in the C_{AlN} with increasing K is the increase in the amount of activated nitrogen species during nitriding reaction.

In the case of P change with constant K, the lifetime of the activated nitrogen species should be shortened with the increase in P, because the collision frequency of nitrogen molecules and activated nitrogen species is increased with increasing P. On the other hand, the amount of the activated nitrogen species generated by arc discharge is same amount, because E_a is approximately constant. However, the amount of activated nitrogen species during nitriding reaction is decreased with decreasing the lifetime of the activated nitrogen species. Therefore, it is considered that the cause of the decrease in the C_{AlN} with increasing P is the decrease in the amount of the activated nitrogen species during nitriding reaction.

In the case of P_{N2} change with K and P constant, it is considered that the cause of the decrease in the C_{AlN} with increasing P_{N2} is the increase in the amount of the activated nitrogen species which are increased with increasing P_{N2}. Additionally, from the experimental results of C_{AlN} as shown in Fig.5 (c), a necessary the amount of the activated nitrogen species for the nitriding reaction is enough in about $P_{N2}=2$kPa at $K=289$ and $P=10$kPa.

In the present experiments, it is clarified that the parameter of K and P are very important factor for AlN nanosized powder synthesis and/or nitriding reaction. The large amount and long lifetime of activated nitrogen species are necessary for the synthesis of high-purity AlN nanosized powder by PWD in N_2 gas, and these are possible by the optimization of powder synthesis conditions of K and P.

The traditional vapor phase processes are complicated by using NH_3 gas for high purity AlN nanosized powder synthesis and a large-scale apparatus is needed for high mass productivity of AlN nanosized powder[4-8]. In addition, PWD has significant advantages, including high mass productivity for nanosized powder preparation compared with other vapor phase processes. The particle size of AlN nanosized powder synthesized by PWD is smaller than those by other processes, because the particle size of nanosized powder synthesized by the traditional vapor phase process is in the range of 10 nm to several tens nanometer[4-8]. Therefore, it is considered that the PWD process without NH_3 can be synthesize a large amount of high purity AlN nanosized powder with particle size less than 10 nm by using simple apparatus.

CONCLUSION

Synthesis of AlN and Al mixed nanosized powders was carried out using PWD with Al wire at various relative energy (K), pressure of nitrogen gas (P) and nitrogen partial pressure of nitrogen /argon mixed gas (P_{N2}), and the following conclusions were made. In the case of K change with P constant, the AlN content (C_{AlN}) was increased from 2 to 97wt% with increasing K from 2.7 to 289. In the case of P change with K constant, the C_{AlN} was decreased from 97 to 26wt% with increasing P from 10 to 100kPa. In the case of P_{N2} change, the C_{AlN} was decreased from 97 to 0wt% with decreasing P_{N2} from 10 to 0kPa. The highest C_{AlN} of 97wt% with the median particle diameter of 6.3nm was obtained at $K=289$ and $P=10$kPa in nitrogen gas. From these results, it was clarified that high K and low P are necessary for the synthesis of high-purity AlN nanosized powder by PWD in nitrogen gas.

ACKNOWLEDGMENT

This study was supported by Research Fellowships of the Japan Society for the Promotion of Science for Young Scientists and Japan Science and Technology Agency.

REFERENCES
[1]A. W. weimer, G. A. Cochran, G. A. Eisman, J. P. Henly, B. D. Hook, L. K. Mills, T. A. Guiton, A. K. Knudsen, N. R. Nicholas, J. E. Volmering, and W. G. Moore, Rapid Process for Manufacturing Aluminum Nitride Powder, *J. Am. Ceram. Soc.*, **77** [1],3-18 (1994).

[2]Y. G. Gao, X. L. Chen, Y. C. Lan, J. Y. Li, Y. P. Xu, T. Xu, Q. L. Liu, and J. K. Liang, Blue Emission and Raman Scattering Spectrum from AlN Nanocrystalline Powders, *J. Cryst. Growth*, **213**, 198-202 (2000).

[3]E. Panthieu, P. Grange, B. Delmon, Proposal of a composition model for commercial AlN powders, *J. Eur. Ceram. Soc.* **8**, 233-241 (1991).

[4]C. Grigoriu, M. Hirai, K. Nishimura, W. Jiang, and K. Yatsui, Synthesis of Nanosized Aluminum Nitride Powders by Pulsed Laser Ablation, *J. Am. Ceram. Soc.*, **83**, 2631-2633 (2000).

[5]Q. Zhu, W. Jiang, and K. Yatsui, Numerical and Experimental Studies on Synthesis of Ultrafine Nanosize Powders of AlN by Ion Beam Evaporation, *J. Appl. Phys.*, **86**, 5276-5285 (1999).

[6]S. Yu, D. Li, H. Sun, H. Li, H. Yang, and G. Zou, Microanalysis of Single-Phase AlN Nanocrystals and AlN-Al Nanocomposites Prepared by DC Arc-Discharge, *J. Crystal Growth*, **183**, 284-288 (1998).

[7]K. Baba, N. Shohata, and M. Yonezawa, Synthesis and Properties of Ultrafine AlN Powder by Rf Plasma, Appl. Phys. Lett., **54** (23), 2309-2311 (1989).

[8]M. Iwata, K. Adachi, S. Furukawa, and T. Amakawa, Synthesis of Purified AlN Nano Powder by Transferred Type Arc Plasma, *J. Phys. D*, **37**, 1041-1047 (2004).

[9]M. Faraday, Experimental Relations of Gold (and other Metals) to Light, *Philosophical Transactions of the Royal Society of London*, **147**,145 (1857).

[10]F. G. Karioris, and B. R. Fish, An Exploding Wire Aerosol Generator, *J. Colloid Sci.*, **17**, 155-161 (1962).

[11]M. Umakoshi, H. Ito, and A. Kato, Preparation of TiO_2 Particles by Means of Wire Exploding Method, *Yogyo-Kyokai-Shi*, **95**, 124-129 (1987).

[12]V. Ivanov, Y. A. Kotov, O. H. Samatov, R. Böhme, H. U. Karow, and G. Schumacher, Synthesis and Dynamic Compaction of Ceramic Nano Powders by Techniques Based on Electric Pulsed Power, *Nano Structured Materials*, **6**, 287-290 (1995).

[13]W. Jiang, and K. Yatsui, Pulsed Wire Discharge for Nanosized Powder Synthesis, *IEEE Trans. Plasma Sci.*, **26**, 1498-1501 (1998).

[14]Y. Kinemuchi, K. Murai, C. Sangurai, C. Cho, H. Suematsu, W. Jiang, and K. Yatsui, Nanosize Powders of Aluminum Nitride Synthesized by Pulsed Wire Discharge, *J. Am. Ceram. Soc.*, **86**, 420-424 (2003).

[15]Y. A. Kotov, and O. M. Samatov, Production of Nanometer-Sized AlN Powders by Exploding Wire Method, *Nanostructured Materials*, 12, 119-122 (1999).

[16]H. Suematsu, K. Murai, Y. Tokoi, T. Suzuki, T. Nakayama, W. Jiang, and K. Niihara, Nanosized Powder Preparation with High Energy Conversion Efficiency by Pulsed Wire Discharge, *J. Chinese Ceram. Soc.*, **35**, 939-947 (2007).

[17]C. Sangurai, Y. Kinemuchi, T. Suzuki, W. Jiang, and K. Yatsui, Synthesis of Nanosized Powders of Aluminum Nitride by Pulsed Wire Discharge, *Jpn. J. Appl. Phys.*, **40**, 1070-1072 (2001).

[18]C. Cho, Y. Kinemuchi, H. Suematsu, W. Jiang, and K. Yatsui, Enhancement of Nitridation in Synthesis of Aluminum Nitride Nanosize Powders by Pulsed Wire Discharge, *Jpn. J. Appl. Phys.*, **42**, 1763-1765 (2003).

[19]Yu A. Kotov, Electric Explosion of Wires as a Method for Preparation of Nanosized Powders, *J. Nanoparticle Research*, **5**, 539-550 (2003).

[20]C. Cho, K. Murai, T. Suzuki, W. Jiang, and K. Yatsui, Enhancement of Energy Deposition in Pulsed Wire Discharge for Synthesis of Nanosized Powders, *IEEE Trans. Plasma Sci.*, **32**, 2062-2067 (2004).

[21]K. Murai, C. Cho, H. Suematsu, W. Jiang, and K. Yatsui, Particle Size Distribution of Copper Nanosized Powders Prepared by Pulsed Wire Discharge, *IEEJ Trans. FM*, **125**, 39-44 (2005).

[22]M.W. Chase Jr., NIST-JANAF Themochemical Tables, Fourth Edition, J. Phys. Chem. Ref. Data, Monograph 9, 1-1951 (1998).

[23]K. Murai, Y. Tokoi, H. Suematsu, W. Jiang, K. Yatsui, and K. Niihara, Particle Size Controllability of Ambient Gas Species for Copper Nanoparticles Prepared by Pulsed Wire Discharge, *Jpn. J. App. Phys.*, **47** [5], 3726–3730 (2008).

[24]Powder Diffraction File, ICDD International Center for Diffraction Data, Swanson, Tatge. : Al (04-0787).

[25]Powder Diffraction File, ICDD International Center for Diffraction Data, Swanson, Tatge. : AlN (46-1200).

[26]K. Noguchi, N. Okazaki, T. Akutagawa, K. Abe, and S. Hayakawa, A Current Probe for Nitrogen Laser Circuits, *Jpn. J. App. Phys.*, **19**, L585-L587 (2008).

[27]C. Cho, Y. W. Choi, and W. Jiang, Time Resolve Spectroscopic Investigation of an Exploding Cu Wire Process for Nanosized Powder Synthesis, *J. Korean Phys. Soc.*, **47** [6], 987-990 (2005).

DUCTILE DEFORMATION IN ALUMINA/SILICON CARBIDE NANOCOMPOSITES

Houzheng Wu[1] Steve Roberts[2] Brian Derby[3]
[1]Department of Materials, Loughborough University, Leicestershire, LE11 3TU, UK
[2]Department of Materials, Oxford University, Parks Road, Oxford, OX1 3PH, UK
[3]Materials Science Centre, The University of Manchester, Grosvenor Street, Manchester, M1 7HS, UK

ABSTRACT

A transmission electron microscope study on cross sections obtained from ground and polished surfaces has revealed that ductile deformation is dominated by dislocations in alumina/silicon carbide nanocomposites containing 1, 5 and 10 vol% silicon carbide particles, and by twinning in unreinforced alumina. The dispersed silicon carbide particles in alumina/silicon carbide nanocomposites restrict the motion of dislocations. A dislocation pinning model is used to compare the possible mechanisms of deformation in alumina and the nanocomposites. Cr-fluorescence piezospectroscopy has been used to characterise the residual stress levels in the materials studied. The measured broadening of the Al_2O_3/Cr^{3+} fluorescence peak indicates a dislocation density of 7.3 - 9.7 x 10^{16} m^{-2} under the indentations in the nanocomposites, whilst the beneath indentations in alumina is 1-2 orders of magnitude smaller.

INTRODUCTION

It has been nearly two decades since Niihara and Nakahira reported significant improvements in bend strength, up to about 300%, of polycrystalline alumina, and also in fracture toughness, by the addition of 5-10% silicon carbide particles with size < 10^{-6} $m^{1,2}$. These were termed "ceramic nanocomposites", and a range of nanocomposites have been investigated with different ceramics chosen as the matrix and particles of various compounds or metals as dispersants. Among these nanocomposites, the most studied have been those with matrices of alumina polycrystals containing silicon carbide particles as the dispersants. A number of different studies have attempted to validate the claimed improvements in mechanical properties of alumina/silicon carbide nanocomposites, as well as to explore any other characteristics of the nanocomposite structure[3,4,5,6,7,8,9,10,11,12].

Large variations have been reported in the flexural strength and fracture toughness values reported by different workers, and it is now generally agreed that the improvement in bend strength is modest at best, and improvements in fracture toughness are negligible. It appears that the bend strength is largely influenced by the processing factors, rather the nanostructure alone[13,14], and apparent fracture toughness and bend strength improvements can be largely attributed to residual surface compressive stresses in the nanocomposites due to their different response to grinding and polishing than is found with monolithic polycrystalline alumina[3,12].

A transition of fracture mode, in bending or in surface grinding, from intergranular in monolithic alumina to transgranular in the nanocomposites has been consistently found. This genuine "nanocomposite effect" has been observed in nanocomposites where the SiC particle content is as low as 1 vol%, and in materials produced by either hot-pressing or pressureless sintering[10,11]. Further, several studies have noted that it is easier to produce a fine polished surface with the nanocomposites than is achievable with alumina[10,11]. This is believed to be related to the decreased incidence of grain-boundary fracture and grain pull-out in the nanocomposite[15]. The wear resistance of the nanocomposites is generally increased by a factor of 3 to 5 over alumina with similar grain size; this improvement is observed in both erosion and in sliding wear[9,16]. The ease of polishing and the superior wear resistance indicates that contact damage in the near surface region of the nanocomposites is somewhat different from that found in alumina.

Our earlier experimental studies[12] indicated that ductile deformation caused by contact damage plays a key role in the behaviour of alumina/silicon carbide nanocomposites. In this paper, we report complementary analyses of sub-surface deformation in ground alumina/silicon carbide nanocomposites and monolithic alumina ceramics of similar grain size by cross-sectional TEM and Cr-fluorescence spectroscopy measurements.

EXPERIMENTAL DETAILS AND ANALYSIS TECHNIQUE

Full details of the methods of preparation and manufacture of the materials have been set out in detail elsewhere[12] but a brief description follows. Samples of alumina/silicon carbide nanocomposites were fabricated by hot pressing in a graphite die at 1650 to 1680 °C for 1 hr. under a pressure of 20 - 25 MPa under flowing argon. The nanocomposite consisted of an alumina matrix material containing 1-10% by volume submicron SiC particles. Alumina powder, with submicron particle size (AKP53: Sumitomo, Tokyo, Japan), was used as the matrix. The SiC particles were a commercial α-SiC powder (UF 45: Lonza - now H. C. Starck, Goslar, Germany), with a mean particle size of ~90 nm. The hot pressed discs were ground to remove the top surface on both sides with an epoxy resin bonded diamond wheel (grit size 150 μm) to achieve a specimen thickness of about 3 mm. A wheel speed of 1250 rpm, table translation speed of 0.8 ms^{-1} and feed depth of 12.5 μm per pass was used in this process. Specimens with polished surfaces were produced by lapping successively with 25, 8, 3 and 1 μm diamond slurry on a Kent III polishing machine (Kemet International, Maidstone, UK) with Lamplan plates (Kemet International) rotating at about 60 rpm and with an external load of 15 N.

Cross section TEM samples with ground or polished surfaces were made by ion-milling[12]. The TEM observations were carried out in JEOL200CX and JEOL4000 miscroscopes.

A microindenter (Matsuzawa MHT-2, Japan) was used to produce indents on 1 μm – polished surfaces of alumina and alumina / silicon carbide nanocomposites, using a Vickers diamond indenter tip. The load range used was 500g and 1000g, with a hold time of 15 seconds.

Fluorescence spectra were acquired using a Raman microprobe system (Renishaw, Wotton-under-Edge, UK) using a He-Ne (632.8 nm) laser with a beam intensity of about 1.0 mW. Measurements were made using a × 40 microscope objective lens with a numerical aperture of 0.65, giving an approximate beam diameter on the specimen of 2 μm. To minimise any peak shift due to temperature fluctuations, all calibrations and tests were performed in a temperature-controlled room; to eliminate any remaining temperature-induced changes, a characteristic neon line at 14431 cm^{-1} was used as a frequency standard; full operational details are given in a previous publication[17].

The collected data were subsequently analysed with curve-fitting algorithms included in the SpectraCalc software package (Galactic Industries Corp., Salem, NH, USA). The line position and width were identified by simultaneously fitting the R1 and R2 fluorescence peaks with combined Gaussian / Lorentz function. Repetitions of measurements on the same location showed that the standard deviation between such measurements is small enough to be ignored. For all probe positions, therefore, only one measurement was made.

THE INFLUENCE OF SIC PARTICLES ON DUCTILE DEFORMATION IN ALUMINA

The possible slip systems in Al$_2$O$_3$ crystal are summarised by Snow and Heuer[18]. The self-energy of a dislocation is proportional to the square of its Burgers' vector (b^2); the magnitudes of the possible Burgers vectors are in the following order:

$$b_{1/3\langle11\bar20\rangle} < b_{1/3\langle\bar1101\rangle} < b_{1/3\langle\bar1021\rangle} < b_{\langle10\bar10\rangle} < b_{1/3\langle21\bar31\rangle} < b_{1/3\langle\bar1012\rangle}$$

Among these, basal slip is expected to have the lowest critical resolved shear stress.

Two kinds of twin have been confirmed to operate in the plastic deformation of alumina: basal twins and rhombohedral twins. Their crystallographic elements were unambiguously determined as following:

Basal twin:
$$K_1 = (0001) \quad \eta_1 = <10\overline{1}0> \quad K_2 = \{10\overline{1}1\} \quad \eta_2 = <\overline{1}012> \quad s = 0.635$$

Rhombohedral twin:
$$K_1 = (10\overline{1}2) \quad \eta_1 = <\overline{1}012> \quad K_2 = (\overline{1}012) \quad \eta_2 = <10\overline{1}1> \quad s = 0.202$$

where K_1 is the twinning plane, η_1 the twinning direction, K_2 the reciprocal twinning plane, η_2 the reciprocal twinning direction, and s the deformation shear. For the twinning process, Kronberg[19] suggested a complicated synchronous movement of quarter-partial dislocations with $1/3<11\overline{2}0>$ Burgers vectors. This mechanism involves the shear of an oxygen layer by a $1/3<11\overline{2}0>$ quarter-partial, followed by an additional shear of the adjoining aluminium layer by a $1/3<2\overline{1}\overline{1}0>$ quarter partial. The repetition of this sequence of alternate motions on successive planes gives the correct structure for the sheared proportion.

Geipel et al[20] described the rhombohedral twinning process in terms of a cross-slip twinning model involving dissociated dislocations. The perfect $1/3<0\overline{1}11>$ dislocation is imagined to dissociate into a leading partial with Burgers vector of $1/21.9<0\overline{1}11>$ and a trailing partial with a Burgers vector of $(1/3-1/29.5)<0\overline{1}11>$. Under applied stress, the leading partial will glide to form faulted loops and approach the trailing partial from behind, and then recombine on $(01\overline{1}2)$ plane. By the forced cross slip of $1/3<0\overline{1}11>$, the dislocation will dissociate again and form another faulted loop. The continuation of this process leads to the formation of a rhombohedral twin.

It has been demonstrated that all of the main slip systems can be activated under tensile or hydrostatic compression loading at temperatures above the brittle-ductile transition temperature of Al_2O_3, (> ~1100 °C) [21]. Twinning, as a special type of slip, was also observed under uniaxial compression at this temperature regime[22].

At temperatures below the brittle-ductile transition temperature, plastic deformation is difficult, and more complex, varying strongly with loading condition. Dislocation slip is in general activated under hydrostatic compression conditions with significant shear components[23]. Sharp point indentation, scratching, and abrasion, produces such conditions. Twins have been widely found in single or polycrystalline alumina under loading conditions such as grinding or polishing, sliding wearing, indentation or scratch, bending tests, uniaxial or hydrostatic compression, and thermal down-shock. Basal twins have been much more frequently observed than rhombohedral twins in all these cases. The thickness of the twins was found to be smaller than a few μm for basal twins and frequently down to tens of nanometres, and for rhombohedral twins, ranged from about 1 μm to about 50 μm. Resolved shear stresses for twinning were measured as 12.6 MPa between 627°C and 1100°C, and 227 MPa at 350°C[24].

When SiC particles are dispersed inside the alumina grains, these particles can act obstacles to dislocation motion and to twin growth. As twinning can be considered as caused by a specific type of slip processes, we simplify the problem here and only consider basal slip in the following analysis.

Assume the SiC particles are single-sized spheres with a radius of $2r$, and dispersed in an alumina matrix in a simple cubic lattice with a lattice spacing $(2r + L)$, as shown in fig. 1(a), where each SiC particle occupies one lattice point. The gap between two particles, L, has the following relationship with the radius of SiC particles (r) and the volume fraction (f):

$$L = \left[\left(\frac{4\pi}{3f}\right)^{1/3} - 2\right] r \tag{1}$$

In order to bypass SiC particles, dislocation lines assume a curved shape; the maximum shear stress needed to bow a dislocation segment into a circular arc is given by the following equation[25]

$$\Delta \tau = \frac{\mu b}{L} \cos\left(\frac{\phi_c}{2}\right) \qquad (2)$$

where ϕ_c is critical angle for the obstacle to be by-passed by a dislocation, μ is the shear modulus of alumina, and b is the Burgers' vector. For alumina, G = 150 GPa, and for basal slip $b_{1/3<2\bar{1}\bar{1}0>}$ = 0.476 nm. The maximum critical shear stress, taking ϕ_c = 0 (appropriate for hard, impenetrable obstacles), for a dislocation to by-pass the SiC particles is thus:

$$\Delta \tau = \frac{71.4 f^{1/3}}{(1.6 - 2 f^{1/3})r} \quad (GPa) \qquad (3)$$

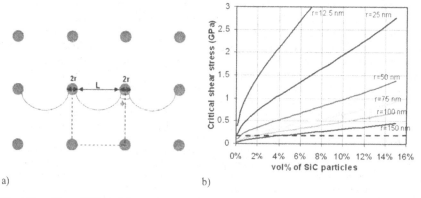

a)   b)

Fig. 1 The effect of SiC particles on the obstacle of dislocation motion in alumina. (a) a simple cubic lattice arrangement of SiC particle with 2r as the diameter of the particles, L the shortest gap between two particles, ϕ_c the critical angle for dislocations to by-pass the particles under an increasing stress. (b) critical resolved shear stress versus the particle size and volume percent, for dislocations to by-pass the particles in alumina/silicon carbide nanocomposites

The modelling results shown in Fig. 1(b) indicates that as little as about 1 vol% SiC particles with a diameter smaller than 150 nm can generate a dislocation motion resistance that is larger than the measured critical shear stress for basal twinning in alumina, 227 MPa at 350°C (represented by the dashed line in fig. 1(b)). For smaller particle sizes, the resistance to dislocation motion increases more rapidly with the increasing volume fraction.

EXPERIMENTAL RESULTS AND DISCUSSION

(a) Ductile Deformation Observed by TEM

For surfaces polished with 1, 3 or 8 µm diamond grits, the deformation induced was dominated by dislocations, as shown in fig. 2. The dislocations were observed to by-pass the SiC particles by bowing, as indicated by arrows. Fig. 1(b) indicates that the level of the resistance dislocation motion for hard particles of this size and spacing is under 1 GPa (particles size >100 nm up to 10 vol %). Compared to the expected level of shear stress around a sharp indenter[26], this level of resistance is not significant; this correlates well with the TEM-based observation that the depths of the dislocated regions underneath the polished surfaces of the nanocomposites and the alumina were very similar at 0.3-1.5 μm[17].

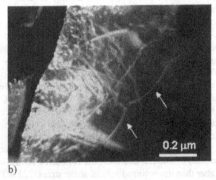

a)   b)

Fig. 2 Cross-section views of the grains underneath polished surfaces of Al₂O₃/SiC nanocomposites. Only dislocations were generated with no twins observed. Dislocations bow around SiC particles, as arrowed.

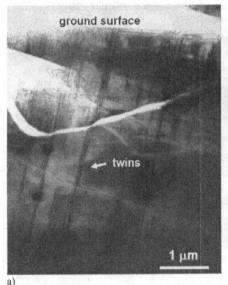
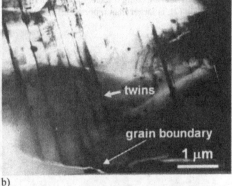

a)   b)

Fig. 3 Cross-section views of ductile deformation beneath the ground surface of polycrystalline alumina. (a) near-surface; (b) near a sub-surface grain boundary.

TEM examination of cross-sections of ground surfaces of unreinforced alumina shows that the ductile deformation is dominated by twins (see Fig. 3). Electron diffraction pattern confirms that these twins are basal twins. The twins initiated on the ground surface, and then grew through the whole grain until meeting the grain boundary; close to the ground surface, dislocations also form, as shown in fig. 3(a), though they do not penetrate as deeply as the basal twins. This confirms that, in pure alumina at low temperatures, twins are easy to generate; the critical shear stress is reported to be ~ 200 MPa at 350 °C[24]. The local temperature during grinding of the ceramic surface is unknown but will be lower than the ductile brittle transition (~1100 °C).

Underneath the ground surfaces of the nanocomposites dislocations dominate, as shown in fig. 4; only occasionally were twins noticed in some grains, but in very low density, as shown in fig. 5. As was found beneath the polished surfaces, high dislocation densities were found either in the deformed surface grains or in grains underneath them up to a depth of 3-10 μm (see previous publication[12]). When twins were developed in a grain of the nanocomposites, their number was significantly reduced, compared to alumina. In fig. 5, only two twins were developed, and the rest of the grain was occupied by dislocations; these twins have by-passed a SiC particle, and become tapered thereafter (see fig. 4(b)). However, neither twin has reached the grain boundary. It is possible that the growth of the twins may have been restricted by the SiC particles in this grain. This implies that the SiC particles impede the propagation of twins in the nanocomposites.

The model predictions in fig 1(b) show that, in alumina dispersed with 5vol% silicon carbide particles of 100 nm diameter, the resistance for the motion of basal dislocations is about 600 MPa, and higher than the required critical shear stress, 225 MPa, for basal twins. The fact that there are few basal twins on the ground surface of Al_2O_3/ 5vol%SiC nanocomposite implies that the simple model is valid, as a first approximation. Accordingly, we hypothesise that, when the resistance generated by the dispersed particles is higher than the required critical shear stress of twins, the growth of twins could be significantly constrained. If this hypotheses is valid, 1vol% of SiC particle could be sufficient to constrain the twinning mechanism in alumina, because the calculated resistance (260MPa), using equation (2), is larger than reported critical shear stress.

Fig. 4 Cross-section view of ductile deformation beneath a ground surface of Al_2O_3/SiC nanocomposite. Note the bowing of dislocation lines between SiC particles, as arrowed.

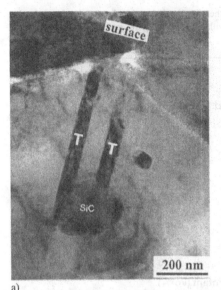

a)   b)

Fig. 5 Cross section view of grains beneath a ground surface of Al_2O_3/SiC nanocomposite. Two twins were initiated at a grain boundary; (a) a SiC particle blocks one of the twins; (b) two twins become thinner on passing a SiC particle.

(b) Dislocation Density

It was not possible to characterize the dislocations, due to the high strains arising from the high dislocation density and from thermal misfit between the alumina matrix and SiC particles. Similarly it was not possible to quantify the dislocation density from TEM images. Hence the broadening of Al_2O_3/Cr^{3+} fluorescence lines was used to estimate the dislocation densities. Details of the methodology are given elsewhere[17].

Sharp indenters, like Vickers indenter, have been used to simulate the grinding process; it is generally agreed that there are similarities in fracture and ductile deformation between the machined ceramics surfaces and the indented surfaces using sharp indenters. Information on the ductile deformation in indents can reflect what could happen on the machined surfaces. By using the measured maximum values of full width at half maximum (FWHM) of the fluorescence peak inside the Vickers indents, we can plot the dislocation density vs the content of SiC particles dispersed inside the nanocomposites, as shown in fig. 6. In this chart, the measurements included those for indents created under different normal loads, i.e. 5 N and 10 N. Under the same indenting load, the nanocomposites have a higher dislocation density than alumina, with a difference of about 1-2 orders of magnitude. This difference echoes the observed difference in the ductile region on the ground surfaces of alumina and the nanocomposites.

These results also firmly demonstrate that there is little difference in dislocation density in the nanocomposites dispersed with different SiC contents. It seems that a dispersion of 1 vol% SiC is enough to have a different ductile deformation in alumina. This ductile deformation should be dominated by dislocations, unlikely to be twinning, otherwise the dislocation density should not be the same as those measurements on the nanocomposites with 5 and 10 vol% SiC particles.

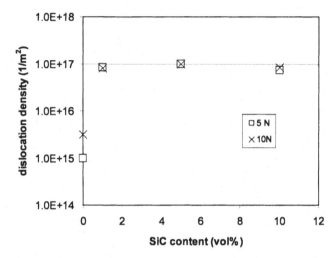

Fig. 6 Dislocation density underneath the indents of Al_2O_3 and Al_2O_3/SiC nanocomposites with different vol% of SiC dispersants. The indents were created with Vickers indenter with a normal load of 5N and 10N.

SUMMARY

Cross sectional TEM examination of the ground and polished surfaces shows that SiC particles dispersed in alumina matrix are likely becoming obstacles to dislocation motion. When the particle sizes are in submicron range, the resistance for dislocations in alumina to by-pass the SiC particles is less than 1 GPa, which is too small, comparing to the shear stress around a sharp indenter, to significantly influence the size of dislocation activating region underneath the machined surfaces or indents. However, it was this resistance that is high enough to constrain the development and growth of basal twins. The modelling prediction and the estimated dislocation density both support that only 1 vol% silicon carbide particles in a size of ~100 nm in diameter used in this study, is enough to constrain the twinning. The estimated dislocation density in Vickers indents using the broadening of fluorescence line is 1-2 orders less in alumina than in the nanocomposites; among the nanocomposites, the dislocation density is the same, between 7.3 to 9.7 x $10^{16}/m^2$, for a silicon carbide reinforcements content of 1, 5 and 10vol%.

REFERENCES

[1] K. Niihara & A. Nakahira: Strengthening of Oxide Ceramics by SiC and Si_3N_4 Dispersions, *Proc. 3rd Int. Symp. on Ceramic Materials and Components for Engines*, edt. by V.J. Tennery. The Am. Ceram. Soc., Westerville, Ohio, pp 919-926 (1988).

[2] K. Niihara: New Design Concept of Structural Ceramics-Ceramics Nanocomposites, The Centennial Issue of the Ceramic Society of Japan, *J. Ceram. Soc. Jpn,* **99**, 974-82 (1991).

[3] J. Zhao, L.C. Stearns, M.P. Harmer, H.M. Chan, G.A. Miller & R.F. Cook: Mechanical Behaviour of Alumina-Silicon Carbide "Nanocomposites", *J. Am. Ceram. Soc.*, **76**[2], 503-510 (1993).

[4] T. Ohji, A. Nakahira, T. Hirano & K. Niihara: Tensile Creep-behaviour of Alumina Silicon-Carbide Nanocomposite, *J. Am. Ceram. Soc.*, **77**[12], 3259-3262 (1994).

[5] R.W. Davidge, R.J. Brook, F. Cambier, M. Poorteman, A. Leriche, D. O'Sullivan, S. Hampshire & T. Kennedy: Fabrication, Properties, and Modelling of Engineering Ceramics Reinforced with Nanoparticles of Silicon Carbide, *Brit. Ceram. Trans.*, **96**, 121-127 (1997).

[6] S. Maensiri & S.G. Roberts: Thermal Shock Resistance of Sintered Alumina/Silicon Carbide Nanocomposites Evaluated by Indentation Techniques, *J. Am. Ceram. Soc,* **85**[8], 1971-1978 (2002).

[7] C.N. Walker, C.E. Borsa, R.I. Todd, R.W. Davidge & R.J. Brook: Fabrication, Characterisation and Properties of Alumina Matrix Nanocomposites, *Br. Ceram. Proc.*, **53**, 249-264 (1994).

[8] M. Sternitzke, E. Dupas, P. Twigg & B. Derby: Surface Mechanical Properties of Alumina Matrix Nanocomposites, *Acta Mater.*, **45**[10], 3963-3973 (1997).

[9] J. Rodriguez, A. Martin, J.Y. Pastor, J. Llorca, J.F. Bartolome & J.S. Moya: Sliding Wear of Alumina/Silicon Carbide Nanocomposites, *J. Am. Ceram. Soc.*, **82**[8], 2252-2254 (1999).

[10] A.J. Winn & R.I. Todd: Microstructural Requirements for Alumina-SiC Nanocomposites , *Brit. Ceram. Trans,* **98**[5], 219-224 (1999).

[11] H. Kara & S.G. Roberts: Polishing Behavior and Surface Quality of Alumina and Alumina/Silicon Carbide Nanocomposites, *J. Am. Ceram. Soc.*, **83**[5], 1219-1225 (2000).

[12] H.Z. Wu, C.W. Lawrence, S.G. Roberts & B. Derby: The Strength of Al_2O_3/SiC Nanocomposites after Grinding and Annealing, *Acta Mater.*, **46**[11], 3839-3848 (1998).

[13] J. Perez-Rigueiro, J.Y. Pastor, J. Llorca, M. Ellices, P. Miranzo & J.S. Moya: Revisiting the Mechanical Behavior of Alumina Silicon Carbide Nanocomposites, *Acta Mater.* **46**[15], 5399-5411 (1998).

[14] L. Carroll, M. Sternotzke & B. Derby: Silicon Carbide Particle Size Effects in Alumina-based Nanocomposites, *Acta Mater.* **44**[11], 4543-4552 (1996).

[15] J.L. Ortiz-Merino & R.I. Todd: Relationship Between Wear Rate, Surface Pullout and Microstructure During Abrasive Wear of Alumina and Alumina/SiC Nanocomposites, *Acta Mater.* **53**[12], 3345-3357 (2005).

[16] H.J. Chen, W.M. Rainforth & W.E. Lee: The Wear Behaviour of Al_2O_3-SiC Ceramic Nanocomposites, *Script Mater.*, **42**[8], 555-560 (2000).

[17] H.Z. Wu, S.G. Roberts & B. Derby: Residual Stress Distributions Around Indentations and Scratches in Polycrystalline Al_2O_3 and Al_2O_3/SiC Nanocomposites Measured Using Fluorescence Probes, *Acta Mater.* **56**[1], 140-149 (2008).

[18] J.D. Snow & A.H. Heuer, Slip Systems in Al_2O_3, *J. Am. Ceram. Soc.*, **56**[3], 153-157 (1973).

[19] M.L. Kronberg: Plastic Deformation of Single Crystals of Sapphire – Basal Slip and Twinning, *Acta Metall.* **5**[9], 507-524 (1957).

[20] T. Geipel, K.P.D. Lagerlof, P. Pirouz & A.H. Heuer: A Zonal Dislocation Mechanism for Rhombohedral Twinning in Sapphire (α-Al_2O_3), *Acta Metall. Mater.*, **42**[4], 1367-1372 (1994).

21 J.B. Watchman Jr & L.H. Maxwell: Plastic Deformation of Ceramic-Oxide Single Crystals *J. Am. Ceram. Soc.* **37**[7], 291-299 (1954).

[22] H. Conrad: Mechanical Behaviour of Sapphire, *J. Am. Ceram. Soc.*, **48**[4], 195 (1965).

[23] H.M. Chan & B.R. Lawn: Indentation Deformation and Fracture of Sapphire *J. Am. Ceram. Soc.*, **71**[1], 29-35 (1988).

[24] W.D. Scott & K.K. Orr: Rohmbohedral Twinning in Alumina, *J. Am. Ceram. Soc.*, **66**[1], 27-32 (1983).

[25] D.J. Green, An Introduction to the Mechanical Properties of Ceramics, Cambridge University Press, pp183 (1998).

[26] E.H. Yoffe: Elastic Stress Fields Caused by Indenting Brittle Materials, *Phil. Mag.* **A** 46[4], 617-628 (1982).

Author Index

Author Index

Printed in the United States
By Bookmasters